WILD ANIMALS
OF THE BRITISH ISLES

WAYSIDE AND WOODLAND SERIES

WILD ANIMALS
OF THE
BRITISH ISLES

By MAURICE BURTON, D.Sc.

★

WITH 32 COLOURED ILLUSTRATIONS
AND 54 HALF-TONE ILLUSTRATIONS
FROM PHOTOGRAPHS BY
JANE BURTON AND OTHERS
AND NUMEROUS TEXT FIGURES

FREDERICK WARNE & CO. LTD.
LONDON · NEW YORK

First printing 1968
Reprinted 1973

ISBN 0 7232 0924 3

Made and printed in Great Britain by
William Clowes & Sons, Limited, London, & Beccles and Colchester

584.873

CONTENTS

LIST OF PLATES   vii

PREFACE   xi

ACKNOWLEDGMENTS   xiii

INTRODUCTION   xv

INSECT-EATERS (Order Insectivora)
Mole, Shrews and Hedgehog   I

FLYING MAMMALS (Order Chiroptera)
Bats   28

GNAWING ANIMALS
(Order Lagomorpha) *Rabbit and Hares*   58
(Order Rodentia) *Dormice, Voles, Mice, Rats, Squirrels and Coypu*   71

FLESH-EATERS (Order Carnivora)
Fox, Dog, Badger, Otter, Marten, Stoat, Weasel, Mink, Wild Cat and Seals   118

CLOVEN-HOOFED ANIMALS (Order Artiodactyla)
Red Deer, Roe Deer, Fallow Deer, Sika, Muntjac, Water Deer and Goat   161

LIZARDS   174

SNAKES   183

AMPHIBIANS
Newts, Frogs and Toads   194

BOOKS FOR FURTHER READING

Index   217

LIST OF PLATES

Between pages xvi *and* xvii

1 Common shrew hunting among yew berries
   Hedgehog under a crabtree

2 Mole eating an earthworm
   Litter of young hedgehogs
   Water shrew on the bank of a stream

Between pages 6 and 7

3 Water shrew swimming
   Hedgehog rolling up

4 Pipistrelles in their roost
   European rabbit

Between pages 14 and 15

5 Greater horseshoe bats at their roost
   Lesser horseshoe bat at roost

6 Mouse-eared bat
   Whiskered bat in flight

7 Natterer's bat in flight
   Natterer's bat at roost

8 Bechstein's bat in flight
   Bechstein's bat asleep

Between pages 22 and 23

9 Hibernating dormouse
   Edible or fat dormouse

10 Albino Daubenton's bat
   Serotine at roost

Between pages 38 and 39

11 Leisler's bat in flight
   Barbastelle at roost

12 Bank vole attracted to an apple
    Field or meadow vole

Between pages 54 and 55

13 Harvest mouse with ear of corn
    Wood mouse or long-tailed fieldmouse

14 Noctule bats at roost
    Long-eared bat at roost

Between pages 70 and 71

15 Brown hare in marshy habitat
    Brown hare leverets eight days old
    Brown hare

16 Ship rat in granary
    Common rat, also known as brown rat

Between pages 78 and 79

17 Red squirrel
    Weasel in typical attitude of rearing up

18 Water vole on river bank
    Orkney vole
    Runs and burrows of water vole

19 Harvest mouse washing forepaws
    Harvest mouse and nest among wheat

20 Badger and cub leaving set
    Vixen

Between pages 86 and 87

21 Wood mouse washing
    Common dormouse

22 House mouse
    Grey squirrel

Between pages 102 and 103

23 Fox's earth
    Badger's set

24 Year-old dog fox
    Three-month-old fox cub

Between pages 118 and 119

25   Polecat
     Scottish wild cat

26   Otter swimming
     Portrait of an otter

Between pages 134 and 135

27   Stoat in snow
     Weasel and its prey, a wood mouse

28   Red deer stag roaring
     Sika deer

Between pages 142 and 143

29   Day-old roe kid
     Fallow buck in velvet

30   Pine marten
     American mink
     Coypu

31   Young roe buck
     Red stags in velvet

32   Sand lizard on ling
     Slow-worms

Between pages 150 and 151

33   Adder
     Smooth snake

34   Chinese water deer
     Sika deer

Between pages 166 and 167

35   Sand lizard sloughing
     Grass snake
     Two adders, normal and melanistic

36   Smooth newt (male)
     Crested newt (male)

Between pages 182 and 183

37   Common frog
     Female edible frog

38   Palmate newt
     Smooth newt larva
     Marsh frog

Between pages 198 and 199

39   Grey or Atlantic seals
     Common seal

40   Natterjack toad
     Common toad confronted by a grass snake

PREFACE

When the late Edward Step wrote the original text of *Animal Life of the British Isles* in 1921, the number of British mammals, and available knowledge concerning them, was limited—certainly by comparison with the number of birds on the British list and known facts about them. The number of reptiles and amphibians found in the British Isles is still barely a dozen, even including introduced species, so it must have seemed natural and convenient to Step to include them in the same volume as mammals, even if the result produced an apparent incongruity.

It is usual to arrange the vertebrate animals in a series, with birds separating the mammals from the reptiles and amphibians. Nevertheless, we have every reason to believe that mammals, the highest class of vertebrates and therefore the most specialised, have evolved from the group of reptiles called Theromorpha of the Permian and Jurassic periods. Birds also had a reptilian origin but there is no indication that they have ever formed an evolutionary stage between the reptiles and the mammals. Evidence suggests that birds and mammals evolved simultaneously from separate reptilian stocks.

Under the heading, *Wild Animals of the British Isles*, we include mammals, reptiles and amphibians, and are therefore bracketing together the quadrupeds in a manner that is not only convenient but logical. They are the beasts of the field as opposed to the fowl of the air.

Even before 1939 a tremendous interest had developed in the study of birds, and at the end of World War II this increased enormously. The new information tended not only to outstrip that available for mammals, reptiles and amphibians but almost to obscure the fact that these even existed. One of the factors giving ornithology such a predominance was that existence of well organised societies for promoting its study. The bias began to be corrected in 1950 by the foundation of the *British Herpetological Society* for the study of both reptiles and amphibians. Then, in 1954, the *Mammal Society of the British Isles* was founded and from this has since proliferated the *Deer Society* and the *Bat Group*. In addition several authoritative books on mammals have been published, and in the universities researches have been concentrated on such animals as the mole, about which so little had been known until the publication, in 1960, of the book, *The Mole* by Godfrey and Crowcroft. Studies tend to go in vogues and that book seems to have stimulated particular interest in the mole in the last few years. That is why this book devotes more pages to that species than to most others. This type of unbalanced treatment is necessary as interest switches from one kind of animal to another; when this present book comes to be revised again it may be that bats will need the greater attention. Certainly we seem, at the moment, to be little more than on the threshold of a far deeper understanding of these flying mammals.

Any book on British wild animals must emphasise one aspect more than another and inevitably leave something out. We have sought to present sufficient basic matter for this book to be a work of reference, at the same time dwelling more on those aspects of the subject which make for interesting reading. It is a book for those who are likely to see British animals in gardens or on rambles.

ACKNOWLEDGMENTS

The author and publishers are most grateful for the help given by Photo Researchers in the selection of photographs, and also thank Jane Burton for photographs reproduced on Plates 1, 2 (above and below), 3 (above), 4, 9, 10 (below), 12, 13, 15 (centre), 16, 17 (right), 18 (above and below), 19 (right), 20 (left), 21, 22 (left), 23 (above), 24 (left), 25, 26, 27, 28, 29, 31 (below), 32, 33, 34 (above), 35 (above), 36, 37, 38 (left), 39 (right) and 40 (below); Harold Bastin for Plate 2 (left); Neave Parker for Plates 3 (right), 22 (above) and 24 (above); Sdeuard C. Bisserôt for Plates 5, 6, 7, 8, 10 (above), 11, 14, 35 (below), 38 (above and below) and 40 (above); Arne Schmitz for Plate 15 (above); Julius Behnke for Plate 15 (right) and 31 (above); Photo Researchers for Plate 17 (above); Douglas English for Plate 18 (left); Geoffrey Kinns for Plate 19 (below); Ernest G. Neal for Plates 20 (above) and 23 (below); H. W. Silvester for Plate 30 (above, left); Leonard Lee Rue III for Plate 30 (above, right); Russ Kinne for Plate 30 (below); A. E. Mc. R. Pearce for Plate 34 (below); B. N. Douetil for Plate 35 (centre) and Barrie Thomas for Plate 39 (above).

INTRODUCTION

The vertebrates or backboned animals include fishes, amphibians, reptiles, birds and mammals. Here we are concerned with only three of these groups. Birds and fishes are numerous and each has an extensive literature of its own, but the amphibians, reptiles and mammals are relatively few in number, and the readily accessible books devoted to them are also few. The British species of amphibians number six: the common, crested and palmate newts, the common toad and the natterjack, and the common frog. Two further species, the edible and the marsh frogs, have been introduced at various times during the past few decades. The reptiles also number six: three lizards—the slow-worm, the common or viviparous lizard and the sand lizard; and three snakes—the adder or viper, the grass and the smooth snake.

It is more difficult to state the number of mammals. Leading authorities writing in recent years have not agreed on the total recognised, because the standing of some species is in dispute. For example, the St Kilda mouse, which has become extinct, used to be regarded as a distinct species but is now known to have been, at best, a form of the house mouse. The zoological standing of some of the island races of voles and mice is also sometimes disputed. If we exclude subspecies and races, and also the whales, porpoises and dolphins, the total is a little over forty. Even that total can only be approximate and depends largely on the view taken about introduced animals. The house mouse is almost certainly not indigenous, but it has been with us so long that its date of arrival here is unknown. The ship rat reached Britain in the twelfth century, and the common rat six centuries later, but, although such relative newcomers, they have established themselves so firmly that one would not think of excluding them from a list of British mammals. The grey squirrel was deliberately introduced, whereas the house mouse and the two rats made their own way here. Even so, because it has become widespread, the grey squirrel is never excluded from the British list of mammals. The rabbit is a slightly different case. We are fairly certain now that it was first introduced after the Norman Conquest. There are, however, Pleistocene remains which are difficult to distinguish from the bones of the present rabbit. One theory is that the rabbit was living here in the Pleistocene period and was driven south into Mediterranean Europe during the Ice Age. If this were correct, the rabbit would represent a reintroduction of a mammal formerly indigenous. The position of the fallow deer is somewhat in doubt too, for much the same reason. The sika and muntjac are now firmly established and sufficiently familiar to naturalists for some notes about them to be included. The coypu and mink have also settled in the countryside as feral species.

The word 'feral' originally meant any wild animal, but in the seventeenth century it was given the wider meaning of an animal formerly tame, cultivated or domesticated, that had reverted to the wild state. If this definition were followed precisely our list would have to be very much extended, because in these days of private zoos and widespread pet-keeping all manner of animals are constantly becoming feral within this meaning. For the purpose of selecting the species to be included in these pages it is argued that only when escaping animals start to breed in the wild can they truly be called feral. Without question, the domestic cat has become feral. Many otherwise tame individuals become semi-wild, and from their ranks, as well as from cats deliberately abandoned, there has grown up a population that breeds in the wild and would be far greater but for the constant watch kept by gamekeepers. There is a less firm case for speaking of feral dogs in Britain, but there is at least one authentic instance of an Alsatian bitch that went wild in the West Country and produced a litter of pups some time after doing so.

The herd of Chillingham cattle is an example of the reverse action. If we are to believe that they are direct descendants of the aurochs, this herd represents survivors of a wild species now held in captivity.

The decision to exclude marine animals is determined by considerations of space, but our two native seals, the common and Atlantic seals, are included. Over and above these, there remain the Cetacea: whales, porpoises and dolphins. Although even those people living on the coasts see comparatively little of these mammals, and any seen are usually dubbed 'porpoises' or 'dolphins' without attempt at more

PLATE 1   (*above*) Common shrew hunting among yew berries
(*below*) Hedgehog under a crabtree

(*above*) Mole eating an earthworm, holding it down with forepaws

PLATE 2

(*left*) Litter of young hedgehogs

(*below*) Water shrew on the bank of a stream

precise identification, the number of cetaceans recorded from British waters classifies them as the largest single group of mammals in our fauna.

Within the historical period, several notable mammals have ceased to be represented in the wild in the British Isles, although most of them are still found wild on the Continent. The list includes the beaver and the wild boar, the aurochs and the short-horned wild ox, the brown bear and the wolf.

The existing British mammals represent seven orders: the Insectivora (shrews, mole and hedgehog), the Chiroptera (bats), the Carnivora (beasts of prey, including the seals), the Lagomorpha (rabbits and hares), the Rodentia (gnawing animals), the Cetacea (whales and dolphins), and the Ungulata (hoofed animals). These all resemble the reptiles and amphibia in having a many-jointed internal skeleton, a bony framework giving support to a system of powerful muscles. The most important feature of this framework is the long backbone or vertebral column consisting of a number of bony rings jointed together by outgrowths of processes, and held in position by strong ligaments. On the dorsal surface of each of the vertebrae two processes slant inwards to join, thus enclosing a canal in which lies the spinal cord. The vertebral column, for descriptive purposes, is divided into five regions: cervical, dorsal, lumbar, sacral and caudal. The number of vertebrae in each region varies somewhat in the different classes and orders. With few exceptions, the cervical or neck vertebrae of mammals number seven. The dorsal vertebrae, to which the ribs are connected, are about thirteen (extreme numbers are nine and twenty-two); the vertebrae of the lumbar or loin region are usually six or seven, but they vary inversely from those of the dorsal from two to twenty-three. The sacral vertebrae (about five) are, in the adult, fused together into a solid bone of triangular shape (the sacrum). The caudal vertebrae vary from three (in man) to nearly fifty, according to the length of tail.

At the front or anterior end of the vertebral column is the skull, containing the brain and the sense-organs. In mammals this is a bony case made up of plates interlocking by their serrated margins. In the reptiles and other vertebrates it is a more or less open framework. In adult mammals the lower jaw, or mandible, is the only part of the skull that is separate. At its hinder end it articulates with the lower part of the skull, and is held in position by strong ligaments and muscles.

The ribs articulate with the dorsal vertebrae, and are connected by cartilage at their outer ends to the sternum or breastbone, which is formed of a series of united bones in the middle line of the chest or thorax. The shoulder-blades (scapulae) of the fore-limbs lie across the upper ribs, their under surfaces being flat or concave; and the hind-limbs are connected to the sacrum by means of the hip-bones which are united below to form the pelvis. The skeletons of reptiles and amphibians show slight differences from the typical mammalian skeleton.

There is a great diversity in the teeth of the small number of British mammals. In adult mammals the teeth are so intimately connected with the jaw as to appear outgrowths of it. In fact, they originate in the skin covering the jaw, and the most effective part of their structure is derived from the epidermis, the outer layer of the skin. The centre of each tooth contains a pulp, around which is the bone-like dentine with an outer coat of hard, glossy enamel. In the incisors (cutting teeth) of the rodents, the front of the tooth is protected by a thick plate of hard enamel, while the hind face of the tooth consists of dentine only. As the teeth are used, the dentine wears away more quickly than the enamel in front, thus preserving the chisel-like cutting edge. In the grinding teeth (molars), especially characteristic of the ungulates, the enamel is thrown into ridges and tubercules on the upper surface, so that in the action of chewing those in the upper and lower jaws behave towards each other like the upper and nether millstones in a corn mill.

Four forms of teeth are recognisable in mammals: the incisor in the front of the jaw, the canines or 'eye-teeth' next to them which are rounded in cross section and pointed at the tips, and the cheek teeth further back in the mouth which, for purposes of description, are usually separated into pre-molars and molars. In describing the teeth of mammals, a simple formula is adopted which shows at a glance the number of each kind in one side of both upper and lower jaw. Taking the human teeth as an example, it would be expressed in this way:

i (incisors) $\frac{2}{2}$,   c (canines) $\frac{1}{1}$,   p (pre-molars) $\frac{2}{2}$,   m (molars) $\frac{3}{3} = 32$

the upper figures representing the number of each kind in the upper jaw and the lower figures the teeth of the lower jaw, the total being reached by multiplying by two for the two sides of the skull.

In classifying mammals, more than with other animals, importance is placed on the structure of the bones of the skull, and more still on the characters of the teeth. The precise details are not relevant here, but the broad principles can be exemplified. The Insectivora are generally regarded as the most primitive of the placental mammals (that is, those mammals other than the marsupials, or pouch-bearers, of Australia and America). This has been confirmed by the finding, less than ten years ago, of numerous fossils of small shrew-like animals that are older than the remains of any other known mammal. The teeth of insectivores are not strongly specialised and the cheek-teeth bear small cusps on the crowns. They are, therefore, especially suited to a diet of insects, the small incisors and canines for their capture, the cusped cheek-teeth for chewing hard-bodied prey. The fact that, of the British insectivores, the mole subsists largely on earthworms and the hedgehog on slugs, snails and earthworms does nothing to invalidate the idea that their dentition is insectivorous, and both eat insects to some extent.

Bats have a similar dentition, and the British bats do, in fact, eat insects only. They are separated from the Insectivora on account of their extreme specialisation for flying, involving the tremendous modification of the fore-limbs to carry the flying membrane. The similarity in the teeth of Insectivora and Chiroptera helps to explain, nevertheless, why the otherwise highly specialised bats should be placed next to the primitive shrews and moles.

In the early classification of the animal kingdom, the human species was normally put at the head of the table. This custom prevailed until 1945, when Dr G. G. Simpson, the eminent American biologist, suggested that the correct place for man, together with the apes, monkeys and lemurs, was next in ascending order to the Insectivora and Chiroptera. This view is becoming increasingly accepted. Simpson puts forward the view that just as bats have specialised in flight, man and the monkeys have specialised in brain capacity; but from a strictly zoological point of view we and they have non-specialised teeth and skeletons generally. Such a view may be a blow to human conceit but there is much to justify it.

The teeth of the carnivores are more specialised than those of the orders so far discussed, but all the kinds of teeth are still to be found in their jaws: the incisors in front, usually three in number, followed by the canines, the pre-molars and molars, in that order from the front of each half of both jaws to the back of the mouth. The fang-like canines are used for slashing cuts, in offence and defence, and the cheek-teeth (i.e. pre-molars and molars) for slicing flesh. The specialisation is seen in two features especially—the fang-like canines, already mentioned, and the carnassials. One cheek-tooth on each side of both upper and lower jaws, known as the carnassial or flesh tooth, is much enlarged and has a prominent cutting edge. The function of the incisors is difficult to define accurately. It is often said that they are used for seizing prey. Certainly they are used in picking up small prey or small pieces of food. They are also used for grooming, and especially for relieving an itch on the skin.

Although we have said that the nature of the dentition and the structure of the skull are of prime importance in classifying the mammals, the subject is not as simple or straightforward as this statement might lead us to suppose. If we were to judge wholly by the dentition, the rodents might be placed higher in the scale than is the case. Other considerations must be taken into account and, as a consequence, there are good reasons, too technical to be discussed here, for placing the carnivores at a higher level than the rodents. One of these reasons is based on the evidence of fossils, the carnivores arising later in time than the rodents. This is too wide a subject to be dealt with here, and the most we can hope to do is to discuss some of the elementary principles of classification.

For a long time, the rodents were held to include rabbits and hares as well as rats, mice and squirrels. It was recognised that there were considerable differences between the two groups. As a consequence, rabbits and hares were included in a suborder Duplicidentata, of the order Rodentia, and rats, mice and squirrels formed the suborder Simplicidentata. This distinction was based upon the characters of the incisors. Rabbits and hares have two pairs of incisors, one behind the other, in the upper jaw. In the Simplicidentata, there is no doubling of the incisors, which number two only in both upper and lower jaws. Added to this, the incisors of rabbits and hares have enamel on the back and front whereas those of rats, mice and squirrels have enamel on the front only. Because the softer dentine at the back of the tooth wears away more quickly, the teeth of the true rodents have a sharper cutting edge. In recent years, the distinction between the two groups has been made more emphatic by elevating the rabbits and hares to a distinct order, the Lagomorpha.

Nevertheless, the Lagomorpha and the Rodentia have a number of features in common. The cheek-teeth have flattened crowns giving a grinding surface, and there are no canines. A wide gap, known as the diastema, separates the incisors and the cheek-teeth. A special fold of the hairy lips can be drawn in between this gap, cutting off the incisors from the rest of the mouth. Thus, when a rat gnaws through a lead pipe or plasterwork, the chips do not enter the mouth.

The last group of all, that formerly known as the Ungulata, has also undergone a change. What was formerly regarded as a single order of hoofed animals is now split into a number of orders, such as Proboscidea for the elephants, Tylopoda, for the camels, and so on. The only hoofed mammals found truly wild in the British Isles today are deer, which are now classified in the order Artiodactyla, or cloven-hoofed animals. These are regarded as the most highly specialised mammals, largely because the limbs are so modified for running. Instead of having five toes, sometimes reduced to four, as do large numbers of the mammals already discussed, deer and cattle have only two and horses one, the toe bones themselves being very much lengthened.

The teeth in the Artiodactyla include strong incisors for cropping vegetation and flat-crowned cheek-teeth for grinding the fibrous material composing it. In some species of cloven-hoofed animals canines are still found, and the ruminants, which include the deer, have six lower incisors. These, together with the canines, which lie alongside them and are not distinguishable from them, project forward. The front of the upper jaw is toothless and bears a hard pad against which the lower teeth work.

The main feature, distinguishing the mammals from all other animals, is the presence of glands in the skin which in the female secrete milk for the nourishment of the new-born young, the milk usually being conveyed through teats. Only in the whales and in the egg-laying mammals of Australia (duckbill and the spiny anteater) are typical mammae (teats) lacking, the milk being secreted from grooves in the whales and from groups of pores in the skin in the egg-laying mammals. There are, in addition, differences in the structure of the skull and in the articulation of the lower jaw with the skull; the skin is always more or less clothed with hair; the heart has a single left aortic arch; the blood is warm, a character shared with birds only, and the heart and lungs are lodged in a special cavity separated from the abdomen by a muscular partition known as the diaphragm.

It has been said, and with truth, that the possession of a few or many true hairs as outgrowths from pits in the skin is alone sufficient to distinguish a mammal from any other animal. Even a whale, otherwise hairless, may have a few bristles on the snout when young. Although the hairs of mammals may take different forms, they are alike in their origin. The spines of the hedgehog have this same origin, indicating that they are modified hairs. Each hair consists of an outer wall enclosing a central cavity filled with pith, in which is the dark pigment which gives the hair its colour. This pigment is always brown, and the varying tints of the hairs, whether black, brown, tawny, cream-coloured or white, depend upon the amount of pigment and its disposition in the pith, combined with differences in the density of the envelope surrounding it. In some instances, as about the mouth, eyes and ears of the cat, we find particularly long sensitive hairs. These are connected at their roots with the ends of nerves, and are highly sensitive. In the cat, for example, they are probably used by the animal to feel its way in the dark. In seals also these long sensitive hairs, popularly known as whiskers but scientifically known as vibrissae, probably play a very large part when a seal has submerged. This is something that is very difficult to test but we can observe the vibrissae laid back along the side of the face when the seal's head is out of water but brought forward as soon as the head is lowered into the water.

There are no marked colour differences in the fur of the sexes, such as we find in the plumage of birds. On the other hand, male and female may be distinguished by the presence or absence of antlers, as in deer, and in the manes and hair-tufts of some exotic mammals. As a rule, also, the male has a stouter neck and a heavier head. Certain species, such as the Alpine hare and the stoat, undergo a marked seasonal change of colour in the fur in the colder latitudes. This is discussed at the appropriate point in dealing with these species.

Wild animals are ready to flee, even when they are fairly used to the presence of human beings. They also start at the slightest sound or at any sharp movement. These considerations, and the fact that most of our mammals have been fairly thoroughly persecuted, so that they tend to keep very much to cover, leaves us in no doubt that the best way of seeing them in their natural surroundings is by being quiet. One may walk about the countryside looking out for whatever may present itself, but by far the best method, having reconnoitred, is to wait, still and quiet, for the animals to come out. Methods vary much with the kind of animal. I have sat for long periods of time, in a place known to be inhabited by foxes, without seeing a sign of one. My best views of this animal have always been the result of pure chance. On the other hand, smaller mammals, such as mice and voles can be seen more readily by sitting and waiting. It is necessary to know the particular locality, however, as well as to know the habits and habitat of a particular beast.

Some of the species about to be described have a very limited range in our country at present, the red deer, for example, as wild animals, are restricted to the Scottish mountains and glens and the West Country moors, but they or other deer may be studied as tolerably free animals in the New Forest, Epping Forest and in many parks such as those at Windsor and Richmond, as well as in private parks. To the deer we must add the wild cat, the pine marten and the Alpine hare as mammals that must be sought in special restricted areas; but most of the others may be reckoned to be met with, sooner rather than later, in our country rambles.

The same remarks can be made about the reptiles and amphibians. Grass snakes can be found almost everywhere and adders are usually abundant where the ground is sandy and well drained. These remarks may not apply to Scotland although even in that area the chances are that the grass snake is more in evidence than we have been led to believe. By contrast with these two species the smooth snake has a very limited distribution. The similar comparison can be made between the viviparous lizard and the sand lizard, and the common toad and the natter-jack. And when we come to the introduced species of amphibians the distribution has varied so much during the last twenty years or more that it cannot be described briefly.

Since the days of Edward Step, the method of studying mammals has undergone a great change. There is still room for observation of the

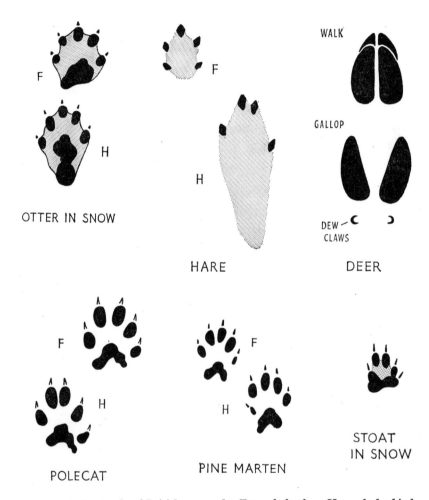

Representative tracks of British mammals: F stands for fore, H stands for hind

kind used by the older naturalist but these are today widely supplemented by live-trapping and by marking. Bats are banded by clipping a small aluminium band to one wing to study their movements, and this has, incidentally, provided statistics of their longevity. Larger mammals, such as deer and seals, can be marked with cattle-tags in the ear. With smaller mammals the most successful method has been to attach to them radio-active rings and to follow their movements with a Geiger counter. The big disadvantage to this is that only one individual can be followed at a time.

The use of live-traps has given valuable information on population densities and on the use of territory. Mammals use a home range, a territory normally used for feeding, and this may be shared with other members of the species. The nest, and a limited area around it, may be defended against intruders of the same species, but there is not the same clear-cut defence of territory at the boundary such as we are familiar

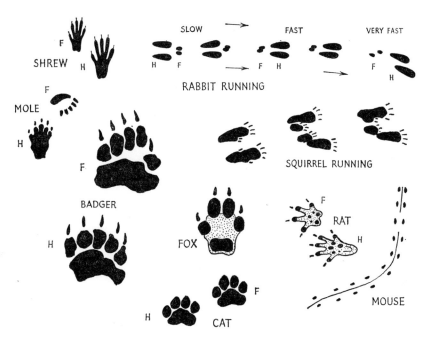

Representative tracks of British mammals: F stands for fore, H stands for hind

with in birds. Under stress of food shortage or other adverse factors the animal may wander beyond the home range. In some species this wandering may be normal in the breeding season. The male of the wild cat, for example, may at such times cover an area much greater than its home range.

One especially interesting fact to emerge from these studies is that there seem to be individuals that are not faithful to a home range. They seem, like the human tramp, to be always on the move. They have been given the name of 'transients'.

INSECT-EATERS

The Mole

Family TALPIDAE   *Talpa europaea*

The mole's body, highly specialised for tunnelling, is cylindrical, with no perceptible neck. The fore-limbs are set well forward, and the hands are relatively large and broad. In addition to the usual five digits, there is a crescentic bone giving greater breadth. The hands are wide open—at most they can be only partially closed—and the palms are always directed face outwards. The bones of the arms, shortened and strengthened, are enclosed within the skin of the body. The strong claws serve as efficient picks to supplement the shovel-like character of the palm.

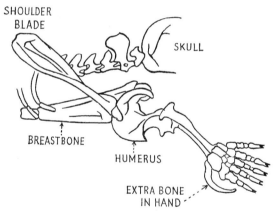

Skeleton of mole's shoulder girdle, including enormously strengthened humerus

The hind-legs are usually described as weak, but this is only by comparison with the powerful fore-limbs. They assist in locomotion, especially when the animal is moving over the surface of the ground, the body moving in a somewhat looping gait, recalling that of a sea-lion. In this, the hind-feet supply the main propulsive force.

The strength of the fore-paws can be best demonstrated by holding a live mole in the closed hand. Very soon its claws are inserted between one's closed fingers, and with an action recalling the beginning of the breast stroke in swimming, the mole's fore-paws force one's fingers apart and the animal's snout is inserted into the opening so formed. This same operation also gives a good demonstration of the normal swimming action of the mole's fore-paws when tunnelling or moving through a permanent tunnel. A similar demonstration is to hold a live mole so that its fore-paws touch the surface of a gravelled path. In a surprisingly short time, the fore-paws will have excavated sufficient of a tunnel to take the head and fore-quarters. In soft earth, the animal appears almost to dive into the ground, and is out of sight in about five seconds.

FORE-FOOT   HIND-FOOT

Underside of fore- and hind-foot of a mole

The skull is unusually long, especially in the nasal bones, and the snout is further carried forward by cartilage lying in the front of these. Externally, the bluntly pointed muzzle ends in a pair of nostrils, set close together. The eyes are fully formed but very small and are hidden in the fur. It seems, however, that the hairs surrounding the eye are under muscular control and can be parted to form a funnel, with the eye at its base. There is little in the way of an external ear, but a firm low ridge can be felt through the fur, which represents all there is of an external ear. Over the snout and head are scattered long bristle-like hairs. These are richly supplied with nerves and blood vessels at the roots, and clearly are sensory.

The dark grey fur covering the body appears black. It is said to be without set: that is, the hairs constituting it arise vertically from the skin and will lie forwards, backwards or to either side with equal ease. This is a great advantage to an animal needing to move backwards or forwards in a tunnel, or even to turn in its own width. Another outstanding feature is that although a mole is habitually travelling underground its fur is never soiled. One secret may be the manner in which it can shake the fur. This is best appreciated if pressure is exerted on the mole's body. A shivering movement passes from the head backwards, sufficiently strong to impart to one's hand the impression of a mild electric shock.

Dimensions vary from one locality to another and from year to year. The males average $5\frac{1}{2}$ in. (140 mm.) head and body, the females being slightly smaller, and in both the club-shaped tail, narrow at the base, is about $1\frac{1}{4}$ in. (32 mm.) long. The weight is at its maximum from February to April and averages nearly 4 oz. (110 gm.) for a male and nearly 3 oz. (85 gm.) for a female. The young reach adult dimensions at two to three months of age.

Colour variations have been recorded; these include normally coloured skins with cream, orange-pink, whitish-black, orange or yellowish markings, as well as skins that are wholly grey, fawn or ash coloured. Albinos have also been found. It is usual, in writing of these colour variations, to say that they occur frequently; and so they do—in museum cabinets, where the number of abnormal skins rivals that of the normal skins. I have questioned mole-catchers on this subject, and the number of skins taken by those to whom I have spoken must, collectively, have totalled more than a hundred thousand. Yet not one of these mole-catchers could recall seeing anything but the normally coloured mole. On the other hand, unusually coloured moles are the subject of news items in the Press from time to time, and from these it seems certain that colour variations tend to be localised, and when one is seen in a district others are likely to crop up there.

Sight is of less importance to a mole than its sense of smell but the latter seems to be less acute than was previously supposed. Its hearing is acute and there seems to be an ability to appreciate vibrations through the ground, as exemplified in the Shakespearian injunction to tread softly, although Shakespeare did say 'that the blind mole may not hear a footfall'. Much more is now known on this point as a result of very recent investigations.

Surprisingly our former ideas about the acuteness of a mole's sense of smell have not been upheld by close observation. On the surface, a mole will search with its nose in a random manner for an earthworm and will fail to locate it until the nose has come within about an inch of it. By contrast we are told that a mole will dig straight up for baby mice lying on the surface. No doubt, there is something of a stethoscopic effect as the odour travels down through the earth, its molecules not being dispersed as freely as in the open air. We can compare this with the way a dog will search with its nose for something lying on the ground and will pick up the odour of truffles growing two feet down in the earth so accurately that its owner knows exactly where to dig.

A similar stethoscopic effect is probably at work with a mole's sense of hearing. The ear can be closed by a sphincter muscle, which probably serves to stop loose grains of earth from entering. Possibly, the mole has the habit of opening and shutting the ear regularly as an emu can be seen to do. It may well be that this is a behavioural trick more common among birds and mammals than is generally suspected at present. So far as the mole is concerned, this ability to close the ear has raised the question of whether it uses the echo-location effect, based on the scratching sounds of its own claws, etc. as it moves about underground, or whether it may use ultrasonics as bats do. The latter is doubtful, although we know the voice is a high-pitched squeal, and this could represent the lower range of ultrasonics. Moles are credited with a twittering noise which seems to be a fast sniffing action. There are also sounds made when it is chewing and the sound of its breathing, which resembles a rapid sniffing as it explores its habitat in the characteristic, feverish manner. These sounds may conceivably aid in a kind of general echo-location, and the hearing may also be helped by such noises being contained by the walls of the tunnel, so that the sound waves are not dispersed as in the open, and thus give a kind of stethoscopic effect.

It has been suggested that a mole may know its way around its tunnels by memory, for example, by remembering that it dug so many paces along this piece of tunnel, then turned right, and so on. Indeed, it is suggested that its restless movements may be linked with a habitual means of committing to memory details of its environment.

A more probable alternative to these methods may eventually be found in the use of the numerous organs of touch with which a mole is endowed. Some of these may be so delicate that they are capable of 'touch at a distance', and a mole's skin is probably supplied with more tactile organs than any other mammal. These consist of thousands of minute papillae on the tip of the snout, known as Eimer's organs, each papilla with its tactile hair, as well as sensitive hairs on the tip of the tail and the so-called Pinker's plates on the skin, especially on the abdomen. These are in addition to the long whiskers on various parts of the head, including the muzzle, which are also highly sensitive.

The mole's body is nearly the same diameter as the tunnel through which it moves and acts like the plunger in a syringe to create air currents underground. Moreover, compression waves in the air between a distant stationary object and a moving mole in a tunnel are probably sufficiently strong and directional to enable its tactile organs, especially the vibrissae, to act as an obstacle detector or a differential distance gauge. The obstacle may be anything from a tunnel blocked by a fall of roof to a trap or a predator. It is by no means uncommon for a gin trap, set in a mole's run, to be jammed with earth, or even a large stone, and on such occasions one finds that the mole has made a detour to reach the other side of the trap.

It was long ago shown that a mole's eye, although tiny, is perfect in construction. Far from being blind there is a very good chance, or so it is suggested, that a mole can see stationary objects but that it takes a perceptible period of time for the retina to register (i.e. the animal is slow to see). More probably it uses its eyes to detect daylight filtering through the soil, and this enables it to tunnel just below the surface.

Everyone must be familiar with the diagrams of what was styled fancifully the mole's 'fortress'. There is, in fact, no more reason for calling it a fortress than for applying the same term to a rabbit's burrow or a bird's nest. Le Court, the French inventor of the term, whose account was published in 1803, described its interior as having a central chamber surrounded by galleries with connecting passages, a central hall and a bolt-hole communicating with the main run. Plans and elevations were made of these details, and for over a hundred years every writer on the mole reproduced these illustrations without questioning their accuracy. This so-called fortress is not to be confused with the 'mole-heaves', sometimes called molehills, or tumps, thrown up at frequent intervals to get rid of the earth from a newly excavated run. The molehill proper, containing the nest, which Le Court named the 'fortress', is about 1 ft. (30 cm.) high and about 3 ft. (nearly 1 m.) across in any direction. These conspicuous mounds are found partly sheltered by a bush, under a spreading tree, or quite commonly at the base of a hedge. They may sometimes be well out in pasture. The nesting chamber lies deeper than the side runs which the mole uses for hunting purposes.

Anyone who has excavated a few dozen of these mounds soon learns that they are not constructed, as has so often been said, to provide a baffling system of bolt-runs for defensive purposes; their design is incidental to the excavation of the nest cavity and the disposal of the material dug out which causes the formation of a solid dome of earth of considerable thickness above it. At certain times of the year, in autumn especially, it is possible to see the molehills not as grass-covered mounds, but as conspicuous heaps of newly turned soil, standing out among numerous heaves in the fields.

The nesting chamber is oval and about 1 ft. (30 cm.) in longest diameter. The nest itself may be made of grass, leaves or slender twigs, or a mixture of these. If when digging out a nest, we compare the materials used with the vegetation litter at the surface, it is very evident that the mole takes such as is immediately available. If the nest is under an oak tree, it will be of oak leaves; under a hawthorn, of hawthorn leaves, and so on. It is also possible to show that the animal does not protrude more than the fore-part of the body from the exits of the tunnels in collecting the nesting materials.

Although the familiar nesting mound has been described here, it is probable that this does not the represent the normal. In 1950 I reported the finding of two nests, each at a depth of 2 ft. (60 cm.) or so below the surface. Others have since been found by other people. Following my original discovery, I have on a number of occasions endeavoured to estimate the number of moles in a field, from their heaves, and then searched for the nesting mounds that showed at the surface. These were always fewer than one would expect from the estimated number of females in a given area. From such evidence, I would suggest that it is normal for the nesting chamber to be well below the surface. Certainly, this must be so in some localities.

The male and female (boar and sow respectively) appear to associate

(*above*) Water
shrew swimming

PLATE 3

(*right*) Hedgehog
rolling up
(photograph taken
just before rolling
up is complete)

(*left*) Pipis-
trelles in
their roost

PLATE 4

(*below*) Euro-
pean rabbit

only temporarily, the female being polyandrous and constructing her own nest-hill, which is said to be smaller and of more simple plan than the male's nest. The hunting tunnels of the female are always said to be winding as compared with the long straight runs of the male. Although I have examined large numbers of runs I have found nothing to confirm this, and trapping suggests that both sexes use the same tunnels. The mating season is at the end of March and beginning of April, and the young are born five to six weeks later. The number of young in a litter varies from two to seven, the usual being three to four. They are blind, naked and pink at birth, but before the fur begins to grow, at about two weeks of age, the skin darkens to a bluish-slate colour. The eyes open about the twenty-second day, and by then the weight at birth, less than $\frac{1}{10}$ oz. (3·5 gm.), has increased to over 2 oz. (60 gm.). The young leave the nest at about five weeks of age and are sexually mature in the February of the following year. The life-span appears to be three years.

The rate of reproduction is controlled to some extent by resorption of foetuses. This may occur in as many as 20% of all cases.

When the juveniles leave the parent they may cover considerable distances either on the surface or in shallow workings. It is believed they are forced on to the surface by adults already occupying a territory; and it may be that occasional records of marked or ringed moles released and trapped again twenty-four hours later as much as one mile (1·6 km.) away, may be sub-adults moving over the surface.

Moles are said to have been originally inhabitants of woodland and fen. They have readily taken to pasture and arable land. Now that modern agriculture has reduced the extent of hedges—the size of fields being increased from ten acres to something of the order of eighty acres in places—trees, shrubs and bushes have been lost. As these provided sheltered ground from which moles could re-colonise adjacent pasture and arable land, changes in their activities have resulted. The use of tractors must also be upsetting to animals so sensitive to vibrations. Established pastures, well manured by sheep, cattle and horses, with high populations of worms, have diminished in number; and arable land treated with pesticides constitute another detrimental factor. This may explain the reports in recent years of plagues of moles in certain types of gardens, parks and other land, which presumably act as refuges. At

3

the same time, a main factor in these increases in numbers is probably the diminution in the number of mole-catchers.

Natural enemies of moles include tawny and barn owls, rats, stoats, weasels, foxes, badgers, cats and herons. In Eastern Europe moles form a considerable proportion of the food of pine martens. Predation by tawny owls is especially heavy in summer, when the young are dispersing and come above ground. Then they may form nearly 50% of the owls' food.

The speed at which a mole travels underground, or over the surface, has often been greatly exaggerated. Speeds 'as fast as a dog can run', or 'a galloping horse' have been suggested. My own tests suggest between $2\frac{1}{2}$ and 3 m.p.h. above ground and below. Other people's estimates, made more recently, put the mole's speed on the surface at 2 m. (3·2 km.) p.h. A mole is a good swimmer and its speed in water is roughly that of its progress over the surface of the ground.

The mole appears to be plentiful in all parts of England, Wales and Scotland; it has even been found at elevations of 2,700 ft. (823 m.). But it does not occur in Ireland, the Shetlands, Orkneys, Outer Hebrides or the Isle of Man. Elsewhere its range is across Europe (except parts of southern and northern Europe) and Asia (south to Himalayas, north to Altai Mountains but not in Japan).

Its food is mainly earthworms, especially the large *Lumbricus terrestris*, although in suitable districts insects form its principal food. Other animal matter, dead or alive, is taken when available, and there are a few records of plant food having been taken. Sea-foods have been accepted when offered experimentally, and it is claimed that a mole will break open and eat eggs, even of domestic fowl. In non-agricultural land the distribution and the vertical movements of both moles and earthworms are closely linked; in soil rich with earthworms moles are abundant. Damp areas in a meadow will show more molehills than the drier parts, and this may indicate not only that more worms are present but also more insects. A localised patch of ground covered by brushwood or branches, making the soil beneath damp, is likely to be visited by moles and the surface of the ground upheaved in a number of heaps of freshly excavated soil. When the brushwood or branches are removed, so that the soil dries, the moles will move elsewhere, or will work at a deeper level, so that no fresh earth appears at the surface.

Insects taken are almost exclusively leather-jackets, wireworms and cutworms. Earthworm cocoons are also widely eaten. A single mole may consume 40–80 lb. (36 kg.) of food in a year, foraging over $\frac{1}{10}$ acre (390 sq. m.). Moles have no need to drink when feeding on earthworms, which consist of about 85% water.

When eating an earthworm, the mole holds it down, using the fore-paws, its hind-feet set wide and bracing the body.

Examination of long runs of mole tunnels shows that earthworms and insect larvae burrow through the walls. These probably supply much of a mole's food, thus reducing the need to dig for that purpose.

Where moles are established there is an extensive system of runs underground. At about 3–6 in. (8–15 cm.) below the surface is a horizontal system of tunnels forming a square-sided network and more or less vertical shafts to the surface. At about 1 ft. (30 cm.) below this is a second horizontal system, connected by more or less vertical shafts to the upper horizontal system. From the lower layer of tunnels there are others which run somewhat irregularly downwards, to end blindly 3–4 ft (1–1·2 m.) below the surface. In addition to these systems, surface runs are used. These are practically at the surface, so that the soil is heaved up, and the run almost opened as the mole drives its way along with about $\frac{1}{4}$ in. (6 mm.) of soil between its own back and the daylight above. In damp meadows the surface of the field may be thickly marked with a network of such runs. No doubt the system of tunnels will vary in design with the nature of the subsoil. The one described here is for a sandy soil and knowledge of it was obtained largely by chance. An arable field was being opened up as a sand pit. One end of the field had already been excavated to a depth of about 14 ft. (4·3 m.). Each day, on one side, about 1 ft. (30 cm.) of the wall was sliced away manually. Over a period of three weeks it was possible for me to visit the sand pit each evening, to see the system of tunnels being exposed in serial sections, and so reconstruct the pattern of the tunnels as a whole.

One mole will work a tunnel system varying from 220–750 sq. yd., its extent marked by the molehills of earth thrown out. Whether such a system represents a territory in the usual sense is not known. There is some evidence of moles moving out into adjacent areas, and of some tunnels being used as common highways. The evidence is somewhat contradictory but my own findings, from studying the excavations in the wall of the sand pit, suggest that these supposed territories are interlinked. Usually a field may show a dozen groups of molehills widely separated, as if each mole is keeping to a well defined territory. Occasionally a small area of cultivated ground may show molehills so closely packed that it is difficult to walk between them.

When excavating, the hands are used alternately and are also used to push earth on to the surface, not the head, contrary to former belief. This erroneous idea seems to have arisen from the way the snout is protruded through newly loosened earth, as if it were being used to upheave the loose soil. The snout is highly sensitive and it is possible that the animal can vary its sensitivity by increasing or decreasing the flow of blood into it. That could explain the conspicuous redness of the tip of the snout at the moment when a mole pushes it through the top of a molehill and moves it this way and that. In that case, what has previously been interpreted as a shovelling action could be resolved into a means of exploring the environment above the ground, either by sniffing or by some teletactile (touch-at-a-distance) sense, for signs of danger.

From the smooth, impacted walls of the tunnels it seems reasonable to suggest that such systems as these have served as highways to generations of moles, perhaps with each generation opening up new ground but using the old tunnels as highways. Such extensive tunnels must affect considerably the drainage of the land. The turning over of the soil, as it is brought to the surface in the heaves, is also a long-term form of ploughing. Although moles are a nuisance on a well kept lawn, in a kitchen garden or on arable land, they must be a beneficial agent in sustaining the fertility of uncultivated ground.

A mole, like its relatives the shrews, is characterised by a restless activity. Its periodicity of rest and activity, throughout the twenty-four hours of day and night, appears to be four-hourly. In fact, it is more likely an alteration of four and a half hours of activity with three and a half hours of rest. At all events, like the shrews, a mole will die of starvation if kept without food for a few hours. A corollary of this seems to be the habit of storing earthworms, when these are especially abundant.

From time to time people have dug out of the soil thousands of earthworms congregated in one spot. For many years, the discussion raged on the meaning of these hoards. Finally, a Danish zoologist kept moles in captivity and supplied them with far more worms than they needed. He found that, under these circumstances, the moles seized one worm after another, bit off the head, twisted the rest into a knot and pushed it into a cavity in the earth. So long as food continued to be sufficient these stores were not touched. In due course, if left alone, the worms grew new heads and could then burrow away. Thus was demonstrated this remarkably effective method of storing food without waste— the uneaten worms lived to be hunted another day.

In the upper jaw there are six incisors of equal size—three on each side—two comparatively large canines of triangular shape and flattened from the sides, eight little pre-molars and six molars. In the lower jaw the dentition is somewhat puzzling, as the canines are similar to the incisors and the first pre-molar works with the upper canine. These teeth are suitable for seizing and chewing insects and other small inverte-

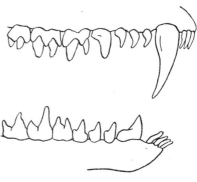

Teeth of mole

brates and agree in general with those of the shrews. The dental formula is:

$$i\,\frac{3}{3}, \quad c\,\frac{1}{1}, \quad p\,\frac{4}{4}, \quad m\,\frac{3}{3} = 44$$

The Common Shrew

Family SORICIDAE   *Sorex araneus*

Along the hedge-bank, the ditch-side or in woodlands, especially where there is long grass, may be seen or heard the high-pitched squeak of one of our smallest mammals. Its actions are restless, and the flexible pointed snout is constantly being twitched and turned, as if exploring the air above. More commonly we find this animal as a forlorn carcase, especially in autumn. It is the common shrew, inoffensive and, from the human point of view, quite harmless. Insects, snails and woodlice, together with carrion and a certain amount of grass seed, constitute its normal diet. A shrew eats about $\frac{3}{4}$ of its own weight of food a day, but this is doubled in a female nursing a litter. A kind of refection has been observed, the shrew everting its rectum to ingest faeces. Surplus food is stored.

The size varies throughout life. The combined length of head and body is about 3 in. (57–76 mm.). The hairy tail is about half this, although the tail length varies a good deal in different individuals. The bilobed snout extends far beyond the mouth, and is abundantly furnished with long whiskers. The hind-foot measures just over $\frac{1}{2}$ in. (15 mm.). The weight averages $\frac{1}{4}$ oz. (7·2 gm.). The body is clothed in a soft coat of close, silky fur, dark brown to almost black on the upper part and paling to dirty yellowish-grey below. The hairy feet are flesh-coloured, and the tail, dark above and light below, is markedly hairy, the hairs being short, stiff and spine-like. Occasionally, the tail may be white-tipped, and sometimes there are white ear-tufts. On each flank is a gland midway between elbow and thigh, the source of the musky odour, a shrew's sole protection against enemies.

There are two moults, the one in autumn leaving a long coat. The moult starts on the hind-quarters and moves towards the head. In spring there is a second moult, which leaves a short coat.

The winter is spent among the leaf litter, but not in sleep as often suggested. In summer shrews move out into the fields and rough pasture, where there are tufts of coarse grass for cover, using runs through the grass, or even tunnels made by mice or moles. Shrews can climb moderately well, using toes that are well separated. They have been seen climbing stout grass stems after insects, and they are said sometimes to climb as much as 8 ft. (2·4 m.) up trees. Although the feet are not particularly well adapted for digging, a shrew can burrow quickly into light or loosened soil, using the fore-feet.

The long, pointed snout is used for turning over dead leaves and loose surface soil in the search for insects, worms and snails; and a shrew's short, soft, velvety fur fits it for passage through the soil without getting dirty.

It is not unusual to see a shrew swimming, and in accordance with this semi-aquatic habit it frequently makes its nest on the banks of ditches. The nursery is a cup-shaped nest woven of dry grass and other herbage with a loose roof beneath which the shrew makes its entrances and exits.

The breeding season extends from May to September in the north, and to October in the south, and during this period each female has at least two litters, each comprising four to eight, or even ten young, the average being just over six. She has only six nipples. The gestation period is not known for certain; the few records available show a range of thirteen to twenty-one days. Resorption of embryos occurs in about 20% of cases. The newly born young weigh $\frac{1}{100}$ oz. (0·5 gm.). The eyes open at eighteen to twenty-one days, which is a couple of days before weaning.

Shrew-flesh is unpalatable and its musky odour repellent to many carnivores, but this does not protect the shrew from death in all cases. Cats, for example, kill many shrews but will not eat one. Owls levy a heavy toll, and other birds of prey, such as the kestrel, are known to take their share, and a further considerable number are taken by magpies, jackdaws, stoats, vipers and smooth snakes.

Where their numbers are high there is a fair chance of casualties through fighting among themselves. There is a general idea that shrews are quarrelsome and always fighting, but in recent years more careful observation of shrews held in captivity, suggests this has been over-stated. It is found that when two shrews approach each other, and their whiskers touch, they squeak. This may result in one of them retreating, usually the intruder, from the territory. Should this not happen, they rear up on to their haunches and continue squeaking, with greater intensity. If at this stage neither gives way, they throw them-selves on to their backs, squeaking and wriggling. It often happens that, as they wriggle, the muzzle of one comes into contact with the tail of the other. This is seized in the teeth. In the course of further wriggling the probability is that the tail of the other is seized in the same way.

Thus we have the spectacle of two shrews in close embrace, squeaking and wriggling; and it was doubtless such incidents that earned these insectivores the reputation of being quarrelsome and always fighting. It is less a fight than a singing contest, although the end achieved is much the same. Injury seldom results, and when it does it is not severe. Because shrews have need of a constant supply of food, overcrowding cannot be tolerated. Except during actual mating, shrews are solitary. The 'singing contests' result in keeping them evenly spaced on the ground so that they make the maximum use of the available food.

Shrews' squeaking contest

In all small animals, owing to the great surface area in proportion to the volume of the body, there is a high loss of body heat. Moreover, the smaller the animal, the more quickly it moves. There is, therefore, a proportionately greater need for food to repair the loss. The more food needed, the greater the need for hunting, and the more hunting the greater the need for food. This vicious circle leads to a pressing need for food at short intervals, so we find the twenty-four hours of the day and night divided, in a shrew's life, into periods of alternate sleeping and feeding, with peaks of feeding approximately every three hours. This, then, is the explanation of the idea, formerly held, that shrews would not live in captivity: they starved in a short while. More precisely, each twenty-four hours is divided into ten periods of activity alternating

(*above*) Greater
horseshoe bats at
their roost in a
cave

PLATE 5

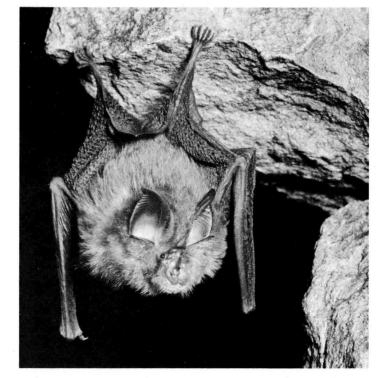

(*right*) Lesser
horseshoe bat at
roost

(*left*) Mouse-eared bat

PLATE 6

(*below*) Whiskered bat in flight

(*above*)
Natterer's bat
in flight

PLATE 7

(*right*)
Natterer's bat
at roost

(*above*)
Bechstein's bat
in flight

PLATE 8

(*left*) Bechstein's
bat asleep

with slightly shorter periods of rest. There are peaks of activity at
10.00 and 20.00 hours, with a drop in activity at about 15.00 hours. Also,
activity is more intense during the night than by day.

Shrews feed mainly below the surface, or in the leaf litter. When seen
above ground they are usually already suffering from undernourishment,
because foraging is less productive for them on the surface than under
cover, and in a short space of time they begin to lose energy. On the
surface, a shrew is, as likely as not, to be seen to totter, or even to drop
dead. So arose the idea that a shrew would fall dead at the sight of a
human being.

There was another belief: that a shrew could not cross a human footpath and live. This may be partly related to the surface starvation,
already noted, but it may also be explainable, at least partially, by
another circumstance. Shrews are short-lived, old age being achieved in
fifteen months. Shrews born this year are reaching the end of their span
in late summer or autumn of next year. There is a heavy mortality in
autumn, but shrews dying underground or in the grass will not be seen.
Only the corpse on the bare earth of a footpath is likely to be seen.

Another idea generally held was that a shrew would readily die from
shock, from the report of a gun or even the bursting near it of a blown-up
paper bag. This, again, was because shrews seen on the surface were
already on the point of death, as a rule, and the explosive sound either
finished them off, or coincided with the moment when they would have
died anyway. Shrews, live-trapped and properly fed can be kept in
captivity without difficulty. As to dying from shock, I have had a pet
shrew jump from the hand, fall several feet on to a concrete floor and,
after being recaptured, continue to live as if nothing untoward had
happened, dying eventually of old age.

The common shrew is found throughout Great Britain but not in
Ireland. It is also absent from the Scillies, Lundy, Man, Outer Hebrides,
some Inner Hebrides (Eigg and Muck), Orkney and Shetland. It is
found in Jersey but not in other Channel Islands. Its vertical range is
from sea-level certainly to 1,500 ft. (450 m.). It is active all the winter
among the dead leaves in some thick hedgerow, where it searches for
hibernating insects which are plentiful in such covers. Nevertheless,
winter shortages do sometimes occur, leading to unusually heavy mor-

tality among shrews. Often the only thing betraying presence of shrews in winter is their shrill squeak, which is not audible to all ears.

A form found in the Isle of Islay was separated as a distinct species under the name of *Sorex granti*, but is now regarded as a subspecies (*S. araneus granti*) of the common shrew.

The shrew's dental formula is: $i \frac{4}{2}$, $c \frac{1}{0}$, $p \frac{2}{1}$, $m \frac{3}{3} = 32$. The summits of the teeth are red-brown, and the almost horizontal lower incisors are encircled by those of the upper jaw.

The Pygmy Shrew

Family SORICIDAE   *Sorex minutus*

The pygmy or lesser shrew is the smallest of all British mammals. It appears to be widely but locally distributed in Britain, the areas in which it occurs being mostly wooded, but extending from sea-level to the tops of our highest mountains. It is found in Ireland, but is not nearly so abundant there as the common shrew is in Britain.

The adult pygmy shrew is only $2\frac{1}{4}$ in. (58 mm.) long. The hind-foot of the common shrew, excluding the claws, is $\frac{1}{2}$ in. (12 mm.), but in the pygmy shrew it is $\frac{1}{6}$ less. The actual length of the tail is about the same in both species. The weight averages $\frac{1}{8}$ oz. (3·5 gm.), with a maximum of $\frac{1}{5}$ oz. (5·6 gm.).

Main distinguishing feature of pygmy shrew is its tail, which is hairy and proportionately longer than that of the common shrew

The colour of the fur is the brown and white of the common species, with a fairly sharp line of demarcation and as in the common shrew there are two moults a year, in spring and autumn. Although the animal as a whole is delicately built, the snout is relatively longer and thicker; the tail is also thicker and more hairy; the fore-arm and hand are shorter. In most of its ways the pygmy is very like its larger relative. The two have much the same habitat and, since the pygmy shrew does not construct runs, it uses those of the common shrew, yet the two seldom mix. Encounters between them are largely avoided because their periods of activity tend not to coincide. For one thing, during any one day the activity and sleep of a pygmy shrew alternate at closer intervals than those of the common shrew. Moreover, the pygmy shrew shows higher peaks of activity during the day while the higher peaks for the common shrew are during the night. When they do meet, the pygmy shrew has the quicker reactions and moves away more swiftly. In any event, fights are restricted to screaming, or avoided by the pygmy shrew burying its head in the ground.

The nests have been found in various situations, such as a clump of rushes, a hollow tree stump or a hollow in the ground roofed by a stone; and they have been of different materials according to the local conditions, moss, dry grass, fine rush shreds and wood chips, variously combined and interwoven to form a hollow ball.

The breeding starts in April, reaches a peak in June and usually ends in August, only rarely going on to October. There are two or more litters a year of from two to eight young.

Outside Britain, the pygmy shrew ranges across Europe and Asia.

Dental formula: $i \frac{4}{2}$, $c \frac{1}{0}$, $p \frac{2}{1}$, $m \frac{3}{3} = 32$.

The Water Shrew

Family SORICIDAE   *Neomys fodiens*

The water shrew is more active by day than by night, and its alternating periods of activity and rest are slightly longer than those of the common shrew. Under water its fur carries a good deal of air trapped in it which gives the submerged body a silvery appearance, 'like an animated air bubble'. On the surface a water shrew chases the whirligig beetles and water gnats; at the bottom it searches for caddis-worms and other larvae. It eats a variety of other aquatic animals, such as snails, worms, small crustaceans, frogs and small fishes; it is not averse to carrion, and has been caught in a trap baited with cheese. Its bite is poisonous. Test have shown that secretion from its submaxillary glands can be lethal even to small rodents. It uses a cricket-like chirp, near the upper limit of the human ear, so that it is audible to young persons but inaudible to the elderly, as with the ultrasonics of bats.

As the water shrew seldom ventures more than a couple of yards from the bank, the observer has a fair chance of taking stock of its activities. Moreover, as the range of its vision is limited, it is not readily alarmed except by sudden noises. It appears to be very buoyant, swimming with the head slightly above the surface and the body apparently flattened dorso-ventrally, but it appears able to walk for a time along the bottom, and at times makes distinct leaps out of the water, presumably after flying insects.

Skeleton of water shrew

The water shrew is larger and has a finer, more dense fur than the other British species of shrew. The length of its head and body combined varies from 3–3¾ in. (76–96 mm.); the body is bulky and the tail longer than the body. The upper parts are dark, from a slaty black to dark brown, and the light ash grey or dirty white of the underparts, which are sharply separated, appear pure white by contrast. The snout is short and broad; the small eyes are blue. The ears are entirely concealed beneath the fur, and each bears a tuft of white hairs. The brown feet are broad and the digits are bordered with stiff hairs which make them very efficient as paddles; the tapering tail of the adult, flattened from side to side, has a double fringe of strong silvery-grey hairs along its underside, constituting a 'keel' and making it more efficient as a rudder. The hind-foot usually exceeds ¾ in. (19 mm.). Weight is about ½ oz. (12–18 gm.) in adults, and is lowest in winter.

Water shrews make shallow burrows in a bank for sleeping quarters, the females burrowing farther in for a nursery, which is a small chamber lined with moss and fine roots. The breeding quarters may, however, be a round nest of woven grass or of leaves. The breeding season is from mid-April to September, with a peak in May or June. The gestation period is twenty-four days. Litters consist of five to eight blind and naked young, each weighing just over 1 gm. These develop rapidly, to become independent at five or six weeks of age. There is probably a second litter in September.

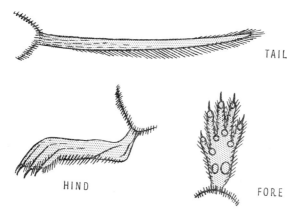

Tail and feet of water shrew. Tail has a keel of hairs and acts as a rudder. Toes are fringed with hairs which assist swimming

There is no hibernation. In winter water shrews may be seen pursuing their prey beneath the ice. Their chief enemies are owls, but they are probably also taken by the predatory mammals which prey on the common shrews; many fall victim to the larger predatory fishes.

The water shrew is much more local in occurrence than the other shrews. With this reservation it may be said to be widely distributed throughout England, Wales and Scotland; in Staffordshire and Cheshire it has been found at elevations of 1,000 ft. (300 m.). It is not found either in Ireland, the Isle of Man, the Outer Hebrides, the Orkneys or Shetlands. In the Fen country it is known as the blind-mouse.

It always comes as a surprise to people who find a water shrew miles from the nearest river or lake, and it is not yet settled whether these are resident in such places or whether their presence there is transitory, just as otters are sometimes encountered miles from water. Normally the water shrew has a home range of up to 60 yd. (55 m.) but live-trapping experiments, made within the last few years, have shown that some mammal species include residents, the majority, which keep to a home range, and transients, those that seem to be more nomadic. It may be that water shrews found far from water are transients, or wanderers.

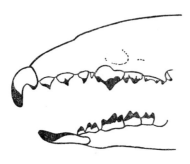

Teeth of water shrew showing dentition typical of shrews

The teeth have coloured tips like those of the other shrews, but the points of the incisors are more hooked than in the two species of *Sorex*; moreover, there are two teeth less, the dental formula standing thus: $i\frac{3}{1}$, $c\frac{1}{1}$, $p\frac{2}{1}$, $m\frac{3}{3}=30$. It is mainly the differences in the dentition that has caused the water shrew to be placed in a separate genus.

The Hedgehog

Family ERINACEIDAE   *Erinaceus europaeus*

The hedgehog, urchin or hedgepig is so distinct from every other British mammal that there is little difficulty in naming it correctly at sight. Its most characteristic feature is the coat of spines, which are modified hairs. There are other peculiarities, such as the extreme shortness of the head and neck in comparison with the bulk of its body, and its ability to roll itself into a tight ball with every part protected by erect spines, and to remain in this position for a considerable time. This is made possible by powerful muscles, which not only draw the coat of spines down at the margins to cover the head and flanks, but are able to maintain the necessary pull without readily suffering fatigue. The first stage in this can be seen when, at first alarm, the hedgehog draws its spines forward over the top of the head. Thereafter, the head, legs and underparts are quickly withdrawn as the edges of the prickly mantle are drawn together to present an almost complete ball of spines.

The best time to see hedgehogs is just before nightfall. Then, one after another can be seen crossing the country roads or the greensward of a common, usually moving quite fast and in a direct line, as if making for a familiar beat. On occasion, however, a hedgehog may be seen abroad in full daylight, after a heavy summer downpour of rain, especially following a dry spell, when the drenched herbage has brought snails and slugs out to feed. These, with insects and worms, constitute the normal diet, but mice, rats, frogs, lizards and snakes may be consumed on occasion. Some vegetable matter, such as acorns and berries, is eaten, and in captivity at least most hedgehogs will eat either apple, pear, orange or other fruit.

A strange and, until recent years, little-known habit of hedgehogs is that of self-anointing. The animal, meeting a substance for the first time, will lick it repeatedly while its mouth becomes filled with a frothy saliva. The substance inducing this may be a cigar end, a handkerchief,

A young hedgehog self-anointing

shoe polish, the skin of a dead toad, a piece of wood and so on. Whatever the substance, when the mouth is full of foam, the hedgehog raises itself on its front legs and throws the head first to one side then to the other, placing flecks of foam on the spines with its tongue. There is, as yet, no satisfactory explanation for this remarkable behaviour.

As a rule, the day is passed in sleep under a heap of dead leaves or moss in a spinney or hedge-bottom. During this time the animal is rolled up but only partially so, in a posture comparable with the coiled attitude of a dog or cat, but it is sufficiently near the fully rolled position for the hedgehog, at the approach of danger, to take up the complete defensive attitude.

It is often said that one may be guided to a hedgehog's sleeping quarters by its snoring. This may be so, but I have never experienced it, nor have I heard it in the hedgehogs I have kept in captivity. A hedgehog not fully rolled, yet apparently aware of an intruder, will snort, or make noises that sound like a gentle snoring, as often as not rocking the body slowly from side to side. What the significance of the 'snoring' may be is not clear. It may be clearing the nostrils to examine better the scent of the intruder. I am inclined to think that this may have given rise to the idea that it snores. Another vocalisation, only heard rarely, is a loud screaming, when the animal is hurt. It seems also to be used as an aggressive call when confronted by a predator, but again only rarely. This screaming is said, also, to be used when a stick is rubbed across a hedgehog's hamstring.

Hibernation covers the period from October to late March or April. Some individuals sleep through the whole period, but in others sleep is intermittent until December or even later, so that it is not unusual to see a hedgehog out and about on frosty nights, or even on snow, until the end of the year. It seems likely that sleep, from October to December, is less profound in younger individuals. The hibernaculum may be a hole in a bank, perhaps one that has been enlarged by a colony of wasps, but more commonly it is the cavity between the buttress roots of a well-grown tree or under a heap of leaf or brush litter, a favourite place being a compost heap. This is lined with dry leaves and moss, carried in by the mouth, and the process appears to be started towards the end of summer. During hibernation the body is nourished by fat accumulated during the summer, but the energy requirements are low, for the temperature of the body drops, breathing is so slight that it can hardly be detected and the pulse-rate drops considerably. Within the body is a dark brown gland which has been named the hibernating gland. It is now spoken of as brown fat, comparable to that found in a similar position, mainly around the neck and shoulders, in many baby mammals, and in some adults, such as bats and rats.

(*right*)
Hibernating
dormouse
accidentally
spilt from
hibernaculum

PLATE 9

(*below*)
Edible or fat
dormouse

PLATE 10 (*above*) Albino Daubenton's bat. Albinism in bats is rare
(*below*) Serotine at roost (Note band on right forearm)

One theory put forward was that the brown fat supplied corpuscles to the blood. Another was that it constituted a ductless gland that poured hormones into the blood. Investigations carried out in 1963 showed the brown fat to be a kind of electric blanket. In baby animals there is a need at birth for extra warmth to replace the protective warmth of the womb. New-born mammals cannot shiver to increase the temperature, which makes necessary this substitute.

Ordinary white fat helps to insulate the body and keep it warm. Such fat is made up of cells each containing only one droplet of fat. The cells of brown fat are not only larger but each contains several droplets of fat. More important, the cells of brown fat contain many mitochondria, whereas the cells of white fat contain only a few. Mitochondria are very small bodies, visible only with high powers of the microscope, that burn up fat to produce heat, and the more there are the more quickly the fat droplets can be converted to heat. The cells of brown fat produce heat twenty times as quickly as white fat. In addition their heat production increases rapidly as the temperature of the surrounding air drops, so this dark coloured fat behaves like a thermostatically controlled electric blanket.

The advantages of the brown fat in the so-called hibernating gland are first, that while the animal is in its winter sleep it produces just enough heat to prevent the hibernant from dying from cold should the temperature of the surrounding air fall, and secondly the more the temperature falls the more the body of the sleeping animal warms up until finally it wakes, becomes active for a while, and so can seek a spot more sheltered from the cold. It is also the brown fat which wakes the hibernating animal up at the end of the winter and warms it so that it quickly becomes active once more.

Other physiological changes associated with hibernation include the laying-in of a store of white fat and action by some of the endocrine or ductless glands. For example, the thyroid, parts of the adrenal glands and the front part of the pituitary gland lying under the brain, undergo a reduction in size and a decrease in their function. The reproductive organs also become reduced in size and the chemical composition of the blood alters rapidly. Over and above these preparations a hedgehog in autumn gives up its ability to control its own temperature, so that it

4

becomes in effect a cold-blooded animal. What this means is not easy to put in a few words, but we can say that changes in the chemistry of the body alter basically the body's reactions to fluctuations in the temperature of the surrounding air. There is also a heavy concentration of the white blood corpuscles in the lining of the stomach and around the main blood vessels, presumably to combat the invasion of bacteria from the intestine and stomach.

Despite the profound sleep of hibernation the hedgehog is still in touch with its surroundings to some extent. It will, for example, respond to sharp clicks by raising its spines slightly at each click. Should the temperature drop too much, the heart, which remains warmer than the tissues on the outside of the body, automatically begins to beat faster. The animal resumes its temperature control, becomes once more warm-blooded and will resume normal activity for a while, afterwards falling asleep again to continue its hibernation.

The hedgehog's eyesight does not appear to be very good, but its senses of smell and hearing are acute. The acuteness of the sense of smell may be gauged by the dampness of the nose, and at times there is a continual succession of drops of fluid falling from the tip of the snout. In other words, its nose is always running. When odours were presented to the nostrils of an anaesthetised hedgehog, electrodes previously inserted into the brain showed that two-thirds of the brain surface was activated. As an example of the acuteness of hearing one may quote the experience of Herter, the German scientist who, in training hedgehogs by 'reward and punishment', used for punishment the clicks made with tongue and teeth usually rendered in writing as 'tut-tut'. Apparently these small, sharp sounds are painful to a hedgehog's sensitive ears.

Hedgehogs have been often condemned for taking hens' eggs, and those of wild birds, including partridges and pheasants. This has now been shown beyond doubt to be true, but eating eggs does not seem to be a general habit. Several people have tested tame or captive hedgehogs and have found that they show no interest in eggs deliberately placed within their reach, or unable to open the egg even if they did show interest. The explanation may be that if a hedgehog finds a broken or cracked egg it then develops a taste for them and behaves accordingly.

Another strong belief among many people is that a hedgehog will take milk from a cow. Certainly it is fond of milk, as everyone knows, but whether it will deliberately help itself is another matter. The usual reason given for disbelieving this is that the gape of its mouth is too small to allow of it getting a cow's dug into it. Anyone who advances this as a reason cannot have seen a hedgehog yawn, or put his finger into a hedge-hog's mouth.

A hedgehog is quite a good swimmer and an even better climber, not only of trees but of drainpipes and rough walls, especially where these are creeper-clad, and there is more than one account of a fight between a rat and a hedgehog in the ivy growing on a wall. In such a contest the two participants seem to be very evenly matched for sometimes the hedgehog wins, and sometimes the rat. There is, however, no doubt that a hedgehog is a match for a viper, against whose poison it is immune. In fact, it is immune to all but a very few poisons, and even then abnormal doses are needed to kill. Gipsies, foxes and badgers appear to be the principal enemies.

The male and female are known respectively as the boar and sow. The peak of the breeding season is between May and July, but there may be second litters during August or September. The gestation period is thirty-one to forty days. Litters consist of three to seven blind, deaf and helpless young, sparsely clad with pale flexible spines, and with the ears drooping. At birth the babies measure $2\frac{1}{4}$–$3\frac{3}{4}$ in. (57–96 mm.) in length and weigh $\frac{1}{3}$–$\frac{2}{3}$ oz. (11–25 gm.). Between thirty-six and sixty hours after birth a second coat of darker spines begins to appear between the spines of the first coat. The ability to roll up does not come until the young are eleven days of age, and three days later the eyes begin to open, one eye opening first, then the other, the process occupying three days. Simultaneously, a third set of spines, ringed with three bands, the middle one dark, the other two lighter in colour, begins to grow out. These are the mature spines, and the first two sets are shed after they have appeared. By this time the baby is a month old, has started to make excursions from the nest, and is being weaned. Growth is rapid: the babies' weight doubles in seven days and has increased tenfold by seven weeks of age. Care of the young is solely by the mother, the male taking no part. The young reach sexual maturity the following year.

The spines are arranged in radiating groups, surrounded by coarse harsh fur. Normally, these lie flat upon the body, but are erected as a reflex to disturbance. They cover the entire upper surface with the exception of the short conical head and stumpy little tail. The head and underside are clothed with harsh fur of a dirty brown or dirty white colour. A hedgehog usually expresses its feelings by means of a quiet grunt; the youngsters give a metallic whistle, almost bird-like. For some days after the spines have hardened the young hedgehog is still unable to roll up. Then, if disturbed, it has the habit of jumping several inches into the air, with painful results to anyone trying to pick it up.

Skeleton of hedgehog

The adult male hedgehog is up to 10¼ in. (260 mm.) head and body, and the tail is a little over 1 in. (25 mm.); the female is less than the male by about ¾ in. (13 mm.). In relation to its entire bulk the neck and body are said to be shorter than in any other British mammal. Both hand and foot have five clawed toes, and five pads on the sole. The weight of an adult may be up to 2⅔ lb. (1·2 kg.), females being lightest in January and heaviest in July, males being lightest in June and heaviest just before hibernation.

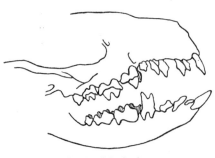

Teeth of hedgehog

The sharply pointed spines are about ¾ in. in length. They are quite hard, and have from twenty-two to twenty-four longitudinal grooves, and a hemispherical base above which is a narrow neck sharply bent, so that the spine is set almost at right angles to the base. The effect of this is that pressure will not drive the spines into the skin, and a hedgehog can fall unharmed from a good height, the spines taking the force of the impact with the ground as they bend at the narrow neck.

When attacked a hedgehog has been known to use the skunk-like habit of emitting a highly objectionable odour, but this does not seem to be a general practice.

The dentition of the hedgehog is: $i\frac{3}{2}$, $c\frac{1}{1}$, $p\frac{3}{2}$, $m\frac{3}{3}=36$.

FLYING MAMMALS

The Bats

Order CHIROPTERA

Apart from birds, bats are the only living vertebrates capable of true flight. Many others, such as flying squirrels and flying fishes, can remain airborne for a limited period by gliding but bats and birds alone among backboned animals use powered flight. There is one obvious modification for flight. Bats have enormously elongated fingers, which are longer than the fore-arm, the middle finger at least being equal in length to the head and body combined. These very long finger bones support a broad web of skin which is continuous with the sides of the body. It also extends to the hind-legs as far as the ankles, and from the hind-legs nearly or quite to the tip of the tail. The web stretching from hind-legs to tail is known as the interfemoral membrane. The upper part of the breastbone is well developed for the attachment of strong pectoral muscles actuating the wings. The collar-bone is strong, giving support to the humerus and the vertebrae of the neck and back are well developed. By contrast, the pelvis and the bones of the hind-legs are weak. This feeble development renders the hind-limbs unfitted for ordinary locomotion. Some bats at least can land on the ground and take off again, or they may run over the ground using the half-folded wings as fore-limbs, for they can do this at a fair speed. Bats that roost in roof-spaces will run over the rafters or over the vertical faces of walls.

The organisation of the bat's brain is considered to be of low order but certain of its senses are very acute. The ears are always well developed, and in many bats they reach a large size. The little lobe that lies beside the entrance to the human ear, known as the tragus, is much elongated in bats and lies in the opening to the ear. It is usual to speak of it then as an earlet. The variation in its shape affords one of the characters for identification of species. Our two horseshoe bats alone are without any prominent tragus.

In flight a bat's manoeuvrability is superior even to that of birds. This is especially evident if we watch the rapidity with which it can change its speed, stopping sharply when in full flight, then making sudden swoops and turning somersaults. More noticeable still is the fact that bats are able to avoid solid objects during flight, even in conditions of total darkness. Not until 1932 was it finally established that this is the result of their use of an echo-location. In this, the bat emits ultrasonic pulses, above the range of the human ear, and the tragus closes the ear at each pulse, opening it when a pulse is ended so that the ear may pick up its echo from a solid object. In the horseshoe bats, and to a greater extent in many non-British bats, the face is ornamented with fleshy flaps known as nose-leaves, which also play a part in echo-location.

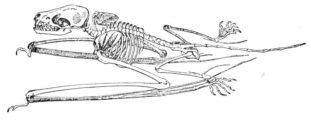

Skeleton of bat

The unravelling of the use made of echo-location constitutes one of the most spectacular discoveries in natural history in the last thirty years. Although bats are not blind, as the popular saying would suggest, nobody knows precisely to what extent they use their eyes. We can be fairly sure that they are not able to use them as a cat does at night. Yet, it has long been known that a bat can fly among the heavy foliage of trees, among all the interlacing twigs, and yet not touch one of them. Formerly it was supposed that some kind of tactile sense in the wing-membranes or in the nose-leaves was responsible. This answer was suggested as far back as 1920 by Professor Hartridge of Cambridge, who failed to pursue his researches further than that. He saw bats flying at night into a room in search of moths. He watched them and noticed that they continued flying from one room to another even when he had switched off all the lights. He then closed a door and found that as long as it was not fully closed, the bat could still fly through. When the door was pushed to a little more, so that the opening was not sufficient to allow passage for it, the bat would fly up to the opening and then fly away without touching either the door or its frame. As a result of these observations he suggested that bats might be capable of detecting the presence of solid objects by means of ultrasonic sounds, that is, sounds above the range of human hearing, reflected into their ears. In 1932 the Dutch zoologist Sven Dijkgraaf proved this was so, and in 1941 two American scientists, D. R. Griffin and R. Galambos, laid the foundations for the very considerable study that has since been developed.

Bats use two kinds of vocalisations. One is audible to the human ear, especially of young persons, and is used as other animals use calls, to express alarm, aggression, and so on. The other is ultrasonic, above the range of human hearing, and is used in echo-location. The first is called a squeak; the second is referred to as a pulse.

While bats are at rest they are emitting pulses at the rate of about ten a second. As soon as they start flying the rate goes up to about thirty a second. Griffin and Galambos arranged a narrow aperture for the bats to fly through and, using an instrument, were able to pick up and measure the pulses they were continually giving out. They found that as the bats came near the hole the pulses became faster and faster and then, when they were safely through, the frequency of the pulses diminished. Later tests showed that the squeaks increase to fifty or sixty a second when the bat is approaching a stationary object and higher than this, up to 200 a second, in some Vespertilionidae, when pursuing an insect. Each pulse lasts approximately 2 milliseconds ($= \frac{1}{500}$ second).

If you shout when standing in a tunnel, the echo comes back quickly. If you are standing in a valley, opposite a nearby hill, the echo takes longer to come back. The nearer you are to the object the more quickly will the echo come back to your ear. If we imagine ourselves moving quickly towards something, the speed at which the echo comes back will increase as the distance between ourselves and the object grows less.

The use of the principle of echo-location is not new in human experience. Formerly skippers of ships crossing the Atlantic would sound the ship's siren to detect the presence of icebergs. The principle has now been expanded for another purpose. By sending out sounds from the ship's hull and then listening for the echoes thrown back from the sea bottom, it is possible to calculate the depth of the sea at that spot. The apparatus used is the depth echo-sounder.

The earliest explanation of why a bat's hearing is not upset while the pulse is being emitted is that the ear is closed by the earlet, or tragus, as the bat emits the pulse. As soon as the pulse has gone out the earlet is raised and the echo comes back and enters the ear. All the time the bat is flying this make-and-break is in operation: a pulse is sent out with the earlet down, then the voice is still while the earlet comes up to allow the echo to enter the ear. A more recent suggestion is that the intensity of the pulses is such that they may deafen the bat, making the fainter echo inaudible. Moreover, pulse and frequency tend to overlap as the rate of pulsing rises. So it is postulated that a bat does not listen for echoes but for the beat frequency, which is audible to the bat even when overlapping occurs.

There are other theories but the most these do is to remind us that we are far from having a complete knowledge of the process. The situation is complicated because different bats have different methods. In horseshoe bats, for example, although the squeaks are emitted through the open mouth, the pulses are given out through the nose. There is no earlet, but instead there is a nose-leaf, the 'horseshoe' from which this bat gets its name. The pulses of a horseshoe bat are given out in almost constant frequency and in bursts, each lasting about $\frac{1}{16}$ of a second, and there are only five or six to a second as against the fifty to two hundred a second of the Vespertilionidae. The sounds are concentrated along a narrow beam and, as a horseshoe bat approaches an obstacle, the margin of the nose-leaf is raised so that the horseshoe becomes somewhat cup-shaped.

It is an interesting coincidence that while Griffin and Galambos were discovering echo-location in Vespertilionidae, Möhres, in Germany, was discovering it in the horseshoe bats. But war conditions made it impossible for the American and German investigators to exchange information or even to know what the other was doing.

It is possible to test the effect of bats' echo-location with a very simple experiment. If a stone is thrown into the air near a flying bat it will turn towards the stone, presumably under the impression that it is an insect. If we throw the stone at the moment the bat is flying overhead and is heading away from us, the bat will turn and fly swiftly back, circle the stone as it drops and go almost down to the level of the stone before it flies away, leaving the stone to fall to the ground. In the brief moment that it almost came in contact with the stone, it somehow knew that the object falling through the air was not edible. Whether this was determined by sight or smell we have not yet discovered. We can, however, test this further by throwing a beetle up into the air instead of a stone; the bat will swoop and take the beetle in its mouth.

These simple experiments also suggest that the bat is not merely measuring the distance of solid objects in front of it but, through its echo-location, is picking up a 'sound picture' of the world around, corresponding to the 'sight picture' that we obtain through our eyes.

British bats are all nocturnal, though a few indulge in occasional flights by day. Most of them have definite hours for flight, the time depending more particularly upon the flight period of the insects upon which they prey. They retire for the day into dark situations, under roofs or in hollow trees, caves or outhouses. In these and similar sleeping places, fair numbers sometimes congregate and more than one species may use the same roost. While asleep their body temperature falls considerably. During bad weather, when the insect prey tends to remain under cover, the bats may not leave their day-time shelter. When the winds are high or it is raining, fewer than normal will leave the roost, the number decreasing in proportion to the deterioration in the weather.

All the British species hibernate, putting on a good deal of fat to carry them through the period of sustained sleep. This sleep is not always as continuous as was formerly supposed, and not only can the occasional bat be seen on the wing in winter, but it is now known that some bats move about within the roost, or go from one roost to another several miles away during the period of hibernation.

The larger bats usually eat their food as they fly, but some of the smaller bats appear to rest at intervals for this purpose. The web between the legs and tail, the interfemoral pouch which, in the female, also serves to receive the newly born young, is often used to catch their prey or to hold it whilst it is being eaten. As with so much concerning the habits of bats, knowledge of the feeding habits must be largely deduced. We may assume that some bats collect prey, and then retire to consume them, from the wings and wing-cases of insects found with accumulations of bat droppings in porches of houses, as well as on the floors of buildings in which bats are known to roost.

Most of the insects bats feed on are small, but a bat's mouth is not very large, so although they may catch some direct with the mouth it would seem that there are auxiliary methods of catching them. A few years ago investigations were made using high-speed photography. Bats were photographed catching tiny fruit flies in a room, mealworms thrown into the air and moths outdoors near an illuminated screen. The smaller insects were mainly caught using the interfemoral membrane as a net and later picked out with the mouth, or were caught directly in the mouth. In addition bats were seen to flex the third and fourth fingers, making a scoop with the wing-tip which directed insects into the mouth or the interfemoral membrane; sometimes a wing would be used actually to flick insects into one or other of these. Another feature revealed by the photographs was that insects are individually located. That is, neither wings nor interfemoral membrane are used purely as dragnets.

The young vespertilionid bat is born blind and naked, the young horseshoe bat has sparse down on its back. The baby bat at once clings to its mother's fur by its claws, and to her nipple by its teeth. Nursing mothers appear to congregate in groups apart from the males, in nursery roosts. The young bat grows rapidly and is soon fully covered with fur. Its eyes open at about seven days. At three weeks of age it takes its first flight but does not lead an independent life until it is about two months old. The longevity of bats was long in doubt, but in recent times, as a result of banding, ages of fifteen years or more have been recorded.

It seems certain that bats, of which there are fourteen distinct kinds in the British Isles, are more numerous in the south of Britain and that there are few species represented in the fauna of Scotland. They also seem to become scarcer towards the west. Most of the species appear to be localised in distribution, the physical features of a district doubtless determining their abundance or scarcity. They appear to be more numerous where there are woods, water, and caves as well as an abundance of insect food. Near local rubbish dumps, for example, which form breeding grounds for many types of insects, the bats, especially serotines and noctules, may congregate by the score within a relatively small area.

The Greater Horseshoe Bat

Family RHINOLOPHIDAE   *Rhinolophus ferrum-equinum*

This is the larger of the two species of horseshoe bat found in the British Isles, the other being known as the lesser horseshoe bat. Horseshoe bats are distinguished by their lack of a tragus and by the conspicuous leaf-like outgrowth of naked skin, known as the nose-leaf, on the muzzle and surrounding the nostrils. The broad fore-part of this has the shape of a horseshoe. There is also a protruding central portion behind the nostrils known as the sella, and behind it is an erect tapering portion known as the lancet. As already mentioned, horseshoe bats emit their pulses, in a narrow beam, through the nostrils instead of through the open mouth, and there can be little doubt that the nose-leaf works in conjunction with this beam, assisting in concentrating the vibration of the pulses into a forward direction. In other words, the nose-leaf acts as a megaphone. Another feature assisting this is the position of the nostrils which are spaced apart by almost exactly half the wave length used by the bat. This spacing results in interference and reinforcement in the sound waves radiating from the nostrils. The effect of this concentration of the emitted sound is that the intensity of the beam declines little with increasing distance. Or to put it simply, the pulses of a horseshoe bat travel much farther but with no greater effort than that of the bats lacking a nose-leaf.

The greater horseshoe bat is heavily built and the female is usually larger than the male. The combined length of head and body is up to $2\frac{3}{4}$ in. (70 mm.) and the length of the tail is $1\frac{1}{4}$ in. (32 mm.). The fore-arm is just over 2 in. (51 mm.) long and the wing-span is up to 15 in. (380 mm.). The weight varies considerably throughout the year, the maximum being in December, when the male weighs up to $\frac{3}{4}$ oz. (23·4 gm.) and the female up to $\frac{4}{5}$ oz. (27·3 gm.). The weight is at its lowest in

April, when there is a decrease in the males of 23% and in the females of 28%. The large ears are about ½ in. (16 mm.) broad, narrowing abruptly to the sharp recurved tip. When laid forward over the face they reach slightly beyond the tip of the muzzle. The lower portion of the broad wing-membrane is attached to the ankle and tail, almost to the tip of the latter. The fur above is thick and woolly, extending a short distance on to both surfaces of the wing-membrane. It is ashy-grey above, pale buff on the underside, often with a pinkish or yellowish tinge. When disturbed its voice is a loud and high-pitched penetrating squeak which some observers have described as a sparrow-like chirp.

This bat emerges from its roost rather late in the evening and it flies at intervals throughout the night. Its flight is usually low, at about 2–3 ft. (60–90 cm.) from the ground, but may be as low as a few inches. The upper limit of its flight is about 10 ft. (3 m.). The action is heavy and butterfly-like, with frequent glides. The greater horseshoe bat feeds on beetles and moths especially, as well as spiders. Small insects are devoured on the wing and the larger prey usually taken to a resting place to be consumed. It has been seen to settle on the ground or on the stems of grass and take ground-dwelling beetles such as the dor beetle. It is also recorded to have been seen on the ground lapping water.

It is gregarious and takes its day-time sleep in summer (May to October) in caves and tunnels, dark buildings, lofts and roofs, and perhaps hollow trees, where it may be found singly but more often in colonies. In summer, at least, the sexes are segregated at the roosts.

Hibernation, which lasts from October to the end of March, is usually in caves and tunnels. The presence of bats overhead is revealed by the heaps of excrement below. The natural resting attitude is hanging downwards by the feet, with the wings wrapped round the body, completely hiding it. Horseshoe bats cannot walk on a flat surface, and before alighting on a vertical one they turn a somersault in the air to get into the proper position. Long ago, Charles Oldham wrote: 'Even when sunk in winter sleep they appreciate a man's approach. Their eyes are, of course, then shrouded by the wings, and the sense of danger must be conveyed to them either by hearing, smell, or, as seems to be most probable, by the exercise of their extraordinary tactile sense, which enables them actually to feel the approaching danger.' Later observers have told how sleeping horseshoe bats will shrink from a finger pointed towards them at close range. We now know that bats are highly temperature-sensitive.

Although the normal hibernation period may be recognised as October to April, it may begin in late September and has been known to continue until May or June, depending on weather conditions. The researches of H. D. Ransome, published in 1967, have shown two things especially. The first is that the behaviour of greater horseshoe bats is governed largely by temperature. The second is that hibernation is certainly not as continuous as has been previously supposed.

In September the females feed heavily and store fat. From the time they go into the hibernaculum until about mid-December they lose weight steadily. Then follows a sharp drop in weight until February, after which their weight begins to increase slightly until they emerge from hibernation. By contrast, the males do not weigh as much as the females when they enter the hibernaculum but their weight remains steady until mid-December when there occurs a drop comparable with that of the females, and there is a similar rise in weight in March or April. A possible explanation for this difference between the sexes is that from late September to mid-December spermatogenesis is at its peak and the males must maintain their energy for this to proceed. Consequently they feed more than the females do during that period.

In the caves which greater horseshoe bats use for their hibernaculum there is a considerable amount of movement. In all the caves used the bats may sleep singly, in small groups of up to half a dozen or in large clusters. To a large extent this distribution is determined by the number of footholds. A cluster, for example, is often along a crack in the rock, where there is a continuous toe-hold. Another feature of the winter quarters is that the bats will move from one cave to another, some taking flights of 20–40 miles (32–64 km.) from one cave to another in winter, or from one part of a cave to another. Banded individuals have been tracked making regular migrations between the same two caves year after year.

The movements within a cave seem to be clearly related to temperature, which varies under the influences of draughts and other factors. From mid-December, at least, until emergence from hibernation the horseshoe bats seek the places in caves where the temperature stands between 52·5 and 56·2°F (6–8°C).

Temperature also influences the sleeping posture. The typical posture, with wings wrapped round the body, is assumed when the bats are at normal temperature but they will cluster, at the same time holding the wings out as the vesper bats normally do, if they need extra warmth. In the summer roosts humidity seems to be the determining influence. When the air is dry the wings are wrapped around the body (typical posture), presumably to reduce loss of body moisture.

During hibernation horseshoe bats feed when the air temperature outside rises to 50°F (10°C). Mr. Ransome has described how he has sat at the mouth of a cave listening with an acoustical tracking apparatus to bats coming out, circling a few times in front of the mouth of the cave, and then going back to sleep when the temperature is below 10°C. With an air temperature of 10°C or more the bats go off to feed, mainly on dung beetles, which appear to be active throughout the winter.

Other evidence of winter feeding Mr. Ransome found was in the guano which accumulates on the floor of the cave beneath sleeping bats. Fresh bat dung becomes quickly covered with a fungal mould that dies down after a fortnight. Moulds on the guano are, therefore, an indicator of freshness, and his examination of the fresh droppings showed they contain wing-cases and other indigestible hard parts of dung beetles.

Mating is promiscuous and takes place in autumn between October and mid-December but the gestation period appears to be only six weeks, although the young are not born until the following June or July. The explanation is that after mating the sperms are quiescent, within a jelly-like plug inside the vagina, until the following spring. This plug is sometimes ejected by the female and can be picked up on the floor of a cave by investigators.

The single baby is born almost hairless, blind, and with pale coloured wings. It has been said to cling to the mother's fur with its claws, and by holding one of the pair of false nipples in her groin, the functional teats being on the mother's chest. It has also been said that the mother at first carries her baby with her when she goes out foraging but that it is not long before she goes out and leaves it hanging from the cave ceiling or the rafters, in a roof space, together with other babies of the colony.

Examination of the roosts when the babies are being born shows newly born babies with the umbilical cord still bleeding, or even with the placenta still attached, hanging by their feet lower down the roof, while the older babies are clustered at the apex of the roof, all this while the mothers are out foraging. As the baby grows older and develops it makes its way progressively up the roof to join the cluster at the apex.

Newly born bats have proportionately large heads, ears and feet. They also have the instinct to climb upwards, so that if they become dislodged they quickly climb back. This is the same instinct that causes the babies in due course to cluster at the apex of the roof, when a roof space is used for summer quarters. Should a baby bat fall to the ground it inevitably lands on or near a heap of guano which is characteristically enveloped in a strong atmosphere of ammonia. This is lethal to the young bat and is probably one of the major reasons why so many baby bats are found dead on the floor of a nursery roost.

Maturity is reached usually at three years of age. The life-span is at least $17\frac{1}{2}$ years.

In this country the greater horseshoe bat is found chiefly in the south of England, especially in the south-west, and in south and west Wales. It does not occur in either Scotland or Ireland. Outside Britain its range extends across Europe and Asia to Japan.

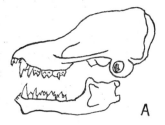 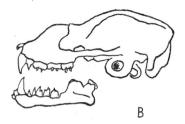

A   Skull of pipistrelle bat showing teeth; B   Skull of greater horseshoe bat showing teeth (the first upper and second lower premolars are minute and cannot be shown here)

The large canine teeth are very conspicuous in contrast with the small incisors. The dental formula for this species is:

$$i\,\frac{1}{2}, \quad c\,\frac{1}{1}, \quad p\,\frac{2}{3}, \quad m\,\frac{3}{3}=32.$$

(*above*) Leisler's
bat in flight

PLATE 11

(*right*) Barbastelle
at roost

(*above*) Bank vole attracted to an apple

PLATE 12

(*left*) Field or meadow vole, also known as short-tailed vole

The Lesser Horseshoe Bat

Family RHINOLOPHIDAE   *Rhinolophus hipposideros*

The lesser horseshoe bat is much smaller and more delicately built than the greater horseshoe bat, which it closely resembles except in size. The nose-leaf has a narrower outline and its sella is more wedge-shaped, the lancet slender with a wedge-shaped tip. The head and body length is 2½ in. (63 mm.) and that of the fore-arm is about the same. The wing-spread is slightly less than 10 in. (254 mm.). The weight—both sexes being about equal in size—is at a maximum of ⅕ oz. (6 gm.) in December and drops to ⅙ oz. (4·9 gm.) in April. The colour is much the same as in the larger species, but a somewhat greyer brown, without the yellow or pink below, and the fur is longer and more silky. Habits also are similar, but naturally, it does not hunt such large beetles, nor does it fly so low— usually between 4 ft. and 15 ft. (1·5–4·5 m.). The lesser horseshoe bat has a more fluttering flight than its larger relative, but it also shows intervals of gliding. The lesser horseshoe is also gregarious, the summer colonies occurring in house and church roofs, in barns and hollow trees; in these the sexes roost separately.

Breeding habits are similar to those of the greater horseshoe. The single young is born in June or July, with a thin coat of downy hair on the upper side only. Sexual maturity may be reached at one year of age. The maximum life-span so far recorded is 12½ years.

Hibernation lasts from October to the beginning of April but is not continuous, the bats often shifting their quarters within the cave, and possibly feeding then on gnats over-wintering in the caves.

Lesser horseshoe bats appear to be more abundant in localities where there are suitable caves, which it shares with the greater horseshoe, but since it has a wider range it is also found without its larger relative. In its winter quarters the lesser horseshoe may hang on the undersides of boulders, a few inches from the ground.

It is most susceptible to wind, and will frequently remain inactive in its shelter when the wind is high outside, although even then one or more of the colony may venture out.

It is a common species in the south of England from Kent to Cornwall, still common, but more sparingly, in Wales, and scarce in Sussex and

5

Hampshire. It is very rare (only three records) in East Anglia, rare in the Midlands, its northward range terminating at Ripon, in Yorkshire. In Ireland it occurs in the west only, in some parts of which it is the commonest species. Outside Britain its range includes Europe south of the Baltic and extends to the Himalayas and North Africa.

Dental formula: $i\frac{1}{2}$, $c\frac{1}{1}$, $p\frac{2}{3}$, $m\frac{3}{3}=32$.

The Whiskered Bat

Family VESPERTILIONIDAE   *Myotis mystacinus*

The small, usually solitary whiskered bat is frequently confused with the pipistrelle although the latter is smaller and has a broader muzzle. The head and body measure up to 2 in. (50 mm.) and the tail is about the same length. The fore-arm is $1\frac{1}{2}$ in. (38 mm.) long. The wings are narrow and have a span of $9\frac{1}{2}$ in. (240 mm.). The weight of a fully grown individual varies around $\frac{1}{6}$ oz. (4·5–6 gm.).

The soft, long fur of the upper parts is a dark or smoky brown, which may be nearly black, and the under surface is lighter, the hairs being tipped with white, but in young individuals the underside also is dark brown. The hair extends but slightly on to the wing membrane, and there is little of it on the long, slender ear, the outer margin of which is deeply notched. The straight, tapering tragus is half the length of the shell of the ear. The hinder margin of the brownish-black wing-membrane is continued to the base of the toes, and the spur (*calcar*) reaches halfway from the ankle to the long tail. The face appears to be very short, owing to the length of the fur on it, and the small eyes are almost hidden. The lips are fringed with long hairs, which suggest its common and scientific names.

The whiskered bat usually rests alone in the day-time, but it has been seen in colonies of a hundred or more in both summer and winter. It makes its appearance early in the evenings, and it may also be seen at times flying in the day-time, especially in spring. Although this bat can be seen at all times during the night it is believed that it hunts inter-mittently rather than throughout the hours of darkness. Flying low, at heights of 4–13 ft. (1·2–3·9 m.), along hedgerows, plantations and cliffs, its normal method of hunting is not to chase flying insects in the air but to pick off those that have settled on leaves and twigs. It likes to be in the neighbourhood of woods and water, where it may take flies, beetles and moths in flight. Its voice is a low, buzzing squeak, but it is quite silent (so far as the human ear can detect) on the wing.

The description of its flight originally given by Edward Step cannot be bettered. 'Individuals did not appear to wander far, but confined their attentions to single pools or short stretches of stream, where they flitted about the alder-bushes or threaded their way with marvellous precision through the lower branches of the sycamore trees. I never saw one rise to a greater height than twenty feet, and often they flew within a few inches of the ground or skimmed the surface of a pool for a yard or two, only to rise again to resume their flight around the alders.' Its flight is slower and steadier than that of the pipistrelle and it has a narrower beat, features which are not easy to appreciate in the half-light.

For its day-time sleep it will shelter in a hollow tree or under a piece of loose bark, a hole in a wall, a roof, or behind window shutters. Hibernation, which lasts from November to March, is in a cave, rock fissure or cellar, the bat emerging for a flight whenever the weather is fine.

On the wing it is not easily distinguished from the pipistrelle, which is similar in size; but the noisiness of the pipistrelle compared with the silence of the whiskered bat is the best guide to those whose hearing is sufficiently acute. It is also more aggressive and ready to bite when being handled, although it can be fairly readily tamed and has been kept in captivity for several months, being fed moths and small beetles.

The single young is born in June or July. The life-span is up to twenty years, but results from the banding of whiskered bats in Holland show that 40% die in the first six months of life and in each year after this there is a 20% mortality. The mean expectation of life is therefore little more than $4\frac{1}{2}$ years.

The whiskered bat is widely distributed throughout England, with the exception of East Anglia, where it is rare. In Yorkshire it has been found at an elevation of 1,400 ft. (425 m.). It appears to be rare in south Wales, but common in north Wales and also in Ireland. It is very rare in Scotland.

There are four other representatives of the genus *Myotis* in the British Isles, the whiskered bat being the smallest member of the genus. All are of slender, delicate form, which is seen most clearly in the shape of the skull, and in the slender muzzle, ear and tragus. They agree also in having thirty-eight teeth—six more than the horseshoe bat. The dental formula of all members of the genus is: $i\frac{2}{3}$, $c\frac{1}{1}$, $p\frac{3}{3}$, $m\frac{3}{3}=38$.

Natterer's Bat

Family VESPERTILIONIDAE   *Myotis nattereri*

The natterer's or red-grey bat is somewhat larger over the spread wings than the whiskered bat, but the head and body measure about the same, up to 2 in. (50 mm.). The fore-arm is $1\frac{1}{2}$ in. (38 mm.) long. The tail is relatively shorter, being only $1\frac{1}{2}$ in. (38 mm.). The wing-span is up to $11\frac{1}{4}$ in. (285 mm.). There are few records of weights, and these, mainly for winter, only show a body weight of around $\frac{1}{4}$ oz. (8–9·5 gm.).

The long fur, soft and dense, is of a greyish-brown colour above and whitish on the underside, with a distinct line of demarcation from the base of the ear to the shoulder. The wing-membranes are dusky. The head is small, with a narrow muzzle which is naked at the tip and slightly overhangs the lower jaw. The face is so densely covered with fur that the small eyes are hidden. There is also a moustache, and above the lips on each side is a prominent gland. The large oval ear is notched on the outer margin above the middle, and the long slender tragus is more than half the length of the ear, ending in a long, very slender point. It has a deep notch near the base on its hind margin. The wing-membrane extends to the base of the outer toe, and the interfemoral membrane is distinctly fringed with stiff hairs along its lower edge. In flight the tail is directed downwards at an angle of sixty degrees or more.

The natterer's bat shares the whiskered bat's partiality for wooded districts. It is both sociable and gregarious, and its day-time retreat in holes in walls, hollow trees, and caverns is shared with bats of its own and other species. It emerges early in the evening and hunts inter- mittently throughout the night, flying at 4–15 ft. (1·2–4·5 m.) from the ground with a slow, steady flight, usually around trees where it often picks flies and small moths off leaves and twigs. It will somersault in the air in order to alight by clinging with its feet.

The single young is born towards the end of June or early in July.

Hibernation ends in March, but there is no record for the time it begins. Surprisingly, this bat is often solitary in hibernation.

Natterer's bat is generally distributed over England, being plentiful in many counties. It also occurs in Wales and is widely distributed in Ireland, but is only occasionally recorded in Scotland. Its total range is from Ireland, across Europe and Asia to Japan.

Dental formula: $i\frac{2}{3}$, $c\frac{1}{1}$, $p\frac{3}{3}$, $m\frac{3}{3} = 38$.

Bechstein's Bat

Family VESPERTILIONIDAE   *Myotis bechsteini*

Bechstein's bat has a general resemblance to the natterer's bat, but is slightly larger, with ears almost twice the breadth of those of that species, and larger feet. The ears are relatively larger than those of any European bat, except the long-eared bat. The wings are narrow and pointed towards the tips.

The body is covered with soft, woolly fur, which is greyish-brown on the upper parts and buff-grey below. The membranes are dark brown; that of the wing reaches to the base of the toes, and that of the inter-femoral leaves the last joint of the tail free.

The combined length of head and body is about 2 in. (50 mm.), of the tail $1\frac{1}{2}$ in.

Large ears of Bechstein's bat showing the earlet (tragus)

(38 mm.). The fore-arm is nearly 1¾ in. (44 mm.) long. The ears are about ¾ in. (19 mm.) in length and ½ in. (13 mm.) wide; the tragus is half the length of the ear. The span of the wings is 10 in. (250 mm.). The few records of weight are for approximately ⅓ oz. (9–11 gm.). The single young is born about midsummer.

Bechstein's bat is the rarest of British bats, and up to 1946 it had been recorded from the south of England only, the localities being the New Forest, Isle of Wight, Sussex, Berkshire and Oxfordshire.

It has since been recorded in Dorset, Somerset, Gloucestershire, Wiltshire and Shropshire. On the Continent it ranges across central and southern Europe, as far as southern Russia, and northwards into southern Sweden.

Our knowledge of its habits is obtained chiefly from the Continent, where it flies about woods, orchards and the neighbourhood of dwellings, coming out from its retreat late in the evening and flying slowly and low over lanes and woodland roads, but only in calm weather.

It lives in small colonies in summer, roosting in hollow trees or in buildings (e.g. in roof spaces).

Its voice is a low buzz, or a high-pitched squeal when excited. It emerges a short while (15 minutes?) after sunset and flies from 4–15 ft. (1·2–4·5 m.) from the ground, feeding mainly on moths taken on the wing or picked off leaves.

Hibernation is in caves but other retreats are almost certainly used, although information on this is lacking.

Dental formula: $i\frac{2}{3}$, $c\frac{1}{1}$, $p\frac{3}{3}$, $m\frac{3}{3}=38$.

The Mouse-eared Bat

Family VESPERTILIONIDAE   *Myotis myotis*

Although this bat has been found in England, it is probably only a rare winter immigrant from the Continent, occasionally becoming temporarily resident. It was authentically recorded for Cambridgeshire in 1888 after having been doubtfully recorded for London before 1835. In 1956 Michael Blackmore discovered several individuals in Dorset, but in 1967 reported that these were no longer to be found. In Europe it has been known to travel long distances, up to 260 km. (162 miles).

It is larger than any of our native species, measuring $2\frac{1}{2}$–3 in. (63–91 mm.) head and body, with a wing-span of 14–18 in. (350–450 mm.). The fore-arm is about $1\frac{1}{4}$ in. (32 mm.) long. The ears are $\frac{3}{4}$ in. (19 mm.) broad and 1 in. (25 mm.) long, and when laid forward extend beyond the muzzle. The mouse-eared bat is medium-brown above and greyish-white beneath, with a clear line of demarcation from the base of the ear to the shoulder. Its roosts are in caves and buildings.

In 1959 the droppings of mouse-eared bats, studied at Bamberg in West Germany, were found to contain considerable remains of non-flying insects and spiders. A. Kolb, who carried out the investigations, observed the bats hopping over the ground, apparently locating dung beetles by smell. They would reject certain leaf beetles but track down a dung beetle in moss, pushing the snout into the moss to capture it. They would similarly dig out a wad of cotton wool soaked with the juice of squashed dung beetles, but would spit it out after chewing it.

Dental formula: i $\frac{2}{3}$, c $\frac{1}{1}$, p $\frac{3}{3}$, m $\frac{3}{3}$ = 38.

Daubenton's Bat

Family VESPERTILIONIDAE   *Myotis daubentoni*

Daubenton's or the water bat was formerly considered one of our rarest bats, but it is now known to be one of the most widely distributed and plentiful, being found wherever there are stretches of water near wooded country. It had probably been mistaken for the common bat or pipistrelle to which it comes near in point of size, though its habits are different. It keeps close to the water, especially to some alder-sheltered pool in the river where there are plenty of caddisflies and other insects. There, from soon after sunset it flies in circles, with a slow quivering flight, frequently dipping its muzzle into the water to pick up surface insects. In such places the evening fly-fisher sometimes finds this bat caught on his hook. It appears to be on the wing all night, but individuals are not continuously active, their night being spent in bursts of activity followed by periods of rest.

It was probably to this bat that Gilbert White referred in his eleventh letter to Pennant, when he said: 'As I was going, some years ago, pretty late, in a boat from Richmond to Sunbury, on a warm summer's evening, I think I saw myriads of bats between the two places; the air swarmed with them along the Thames, so that hundreds were in sight at a time.' This was long before it had been distinguished as a distinct species, and when it would probably have been regarded as the common bat. In summer Daubenton's bat is strongly gregarious at its roosts, in caves, hollow trees, rock fissures and buildings. It emerges late in the evening and up to two hundred have been seen over one lake, hunting for $2\frac{1}{2}$ hours before moving in a body to another lake. It is usually silent on the wing, but gives an angry buzz when alarmed.

The body is clothed with short, dense fur, of a grizzled warm brown colour on the upper parts, and lighter brown or buffy grey, sometimes so pale as to show a distinct line of separation along the sides, from the angle of the lips to the thigh. The face is dusky, and the ears and wing-membrane are of a reddish dusky tint. The interfemoral membrane is whitish below, and there are whitish hairs on the toes. The membrane arises from the middle of the foot.

In size it is a little larger than the whiskered bat and common bat, but smaller than Leisler's bat. The head and body measure about 2 in. (50 mm.), the tail $1\frac{1}{4}$ in. (32 mm.), the ear $\frac{1}{2}$ in. (12 mm.) and the fore-arm $1\frac{1}{4}$ in. (32 mm.): the wing-span is about 10 in. (250 mm.). The few records of weight are for about $\frac{1}{4}$ oz. (7–11 gm.). The fore-leg and foot are conspicuously large. The ear has a rounded tip, and a shallow con-cavity on the upper part of the hind margin; the lance-shaped tragus is about half the length of the ear. The spur, or calcar, of the foot extends three-quarters of the distance between the foot and the tail. The last two joints of the latter usually extend beyond the membrane. In hibernation, which takes place from the end of September to about the middle of April, it is no longer sociable, but hangs alone in some dark cave or in a building, or in groups of up to half a dozen.

There is a single young, born in June or July. The maximum longevity recorded is sixteen years.

Its range extends from Ireland across Europe and Asia to Japan, and from the Mediterranean to central Norway. Throughout the British Isles it is abundant locally.

Dental formula: i $\frac{2}{3}$, c $\frac{1}{1}$, p $\frac{3}{3}$, m $\frac{3}{3}$ = 38.

The Serotine Bat

Family VESPERTILIONIDAE   *Eptesicus serotinus*

The serotine and the noctule are our two largest bats, and in the early records they were very much confused. In both, the wing-span may reach 15 in. (380 mm.). Though similar in size, they may be known apart by the shape of the ear; in the serotine it is oval-triangular, with the tips rounded. The fur is also of a darker brown, and there are other points of difference, such as the possession of two additional teeth by the noctule, and the interior pre-molar being absent from the upper jaw of the serotine. When in flight the much broader wings of the serotine are the best feature for identification.

It has a somewhat swollen face with little hair on the front portion, except for a moustache on the upper lip, but owing to the dark skin of the face the lack of fur is not very noticeable. The dark brown fur, sometimes with a tinge of chestnut, of the upper parts is soft and dense; behind the shoulders the hairs have buffy tips. On the underside the fur is somewhat lighter. There is little extension of fur on the wing, except a line of down on the under surface of the fore-arm. The wing-membrane is attached to the base of the toes. The head and body measure about 3 in. (76 mm.), and the tail slightly exceeds 2 in. (50 mm.), the last joint being quite free of the membrane. The fore-arm measures 2 in. (50 mm.). The wing-span is 13–15 in. (330–380 mm.). The weight is up to 1 oz. (33 gm.), the female being slightly larger. There are prominent glandular swellings on the muzzle. The ear is oval, about $\frac{3}{4}$ in. (19 mm.) long; the short tragus, less than half the length of the ear, has a straight front border and a curved hind border, with rounded tip.

The serotine usually makes its appearance about sunset, but may emerge at any time during the following hour, for a flight of about half an hour. Then it takes a rest and after that hunts intermittently through the night. Typically it frequents glades in woods or wooded parkland and preys upon beetles and moths. It can often be seen over large manure heaps and rubbish pits. In May and June large numbers of cockchafers fall victims to it, and in July and August in Kent and Sussex it plays havoc with the brown-tail moth. It eats on the wing and also at rest; and it may settle on a branch, with wings half spread, to take insects from leaves. Its flight is heavy and fluttering, almost moth-like. In the early part of its season it flies low, at about 8–20 ft. (2·5–6 m.), but later it prefers an altitude between 30 ft. and 40 ft. (9–12 m.), from which, however, it frequently descends to the ground. The change is, no doubt, connected with the seasonal succession of insects with different habits. It is gregarious, and when it retires to holes or roofs for its day-time rest it is usually in companies of up to twenty, but it may roost solitary. Usually silent, it sometimes utters a strident squeak, described as *tick-tick* in flight.

Hibernation is from mid-October until April or May, sometimes mid-March, in hollow trees, in buildings or under loose bark on old or dead trees. Sleep is continuous but the bat has been seen to take wing readily if disturbed.

Serotines in summer roosts are known to cluster before leaving for the night's foraging. They sleep grouped closely together and one by one begin to move about. Then suddenly all will gather in a large cluster on a horizontal surface, such as a beam, or on a vertical surface, such as a chimney breast, vibrating their wings. After this they scatter and one by one fly off in quick succession.

Not a great deal is known about the breeding habits of the serotine. From a very few records indeed we can say that a single young is born in June, perhaps sometimes in May, and that this takes its first flight at the age of three weeks. A specimen found in Surrey in late August was just losing its milk teeth.

Except for a few records of its occurrence in Essex, Cambridgeshire, Suffolk, Hertfordshire, Cornwall and Glamorgan, it might be said to be restricted in Britain to that portion of England bounded by the river Thames and the English Channel, but including the Isle of Wight and Devon, with a few records for Middlesex and Hertfordshire. Even in these places it is only abundant locally. Kent is its main stronghold and there it is the commonest bat. In the Isle of Wight it is known as rattle-mouse.

Outside the British Isles, it extends through central and south Europe, from Denmark to the Mediterranean and eastward into temperate Asia. It also occurs in Tunisia and Algeria.

The canines and the inner incisors of the upper jaw are noticeably large and strong.

Dental formula: i $\frac{2}{3}$, c $\frac{1}{1}$, p $\frac{1}{2}$, m $\frac{3}{3}$ = 32.

Leisler's Bat

Family VESPERTILIONIDAE   *Nyctalus leisleri*

Leisler's or the hairy-winged bat is a smaller edition of the noctule in a darker binding. The length of the head and body is $2\frac{1}{2}$ in. (63 mm.), and of the tail $1\frac{1}{2}$ in. (38 mm.). The fore-arm is about $1\frac{3}{4}$ in. (44 mm.). The wing expanse is 10–12 in. (250–300 mm.). The weight is $\frac{1}{2}$–$\frac{2}{3}$ oz. (14–20 gm.), the females being larger than the males usually. The fur on the upper parts is a darker brown than that of the noctule, but it is lighter on the underparts. The skull is only half the size of that of the noctule, and the Leisler's whole build is lighter and less massive. It does not fly at such heights as the noctule, but more at the level of the pipistrelle, 10–50 ft. (3–15 m.). Its flight is strong but not swift, and its movements are more zig-zag, with occasional shallow dives.

It is one of the rarest of our bats, a woodland species, tending to roost high up in decayed oaks, but also in the roofs and crevices of buildings. Often gregarious, in colonies of up to a hundred, it is at times solitary, especially during hibernation, which lasts from October to mid-April. It is also given to changing its roosts frequently. The evening flight lasts for about an hour, starting just after sunset, and there is a second flight for a similar period which starts just before sunrise. Its food consists of flies, beetles and moths, which it consumes so rapidly on the wing that it returns gorged to its roost.

The single young is born in June.

The distribution of the Leisler's bat does not agree at all with that of the noctule. It has been recorded for Devon, Somerset, Gloucestershire, Hampshire, Surrey, Kent, Essex, Hertfordshire, Cambridgeshire, Norfolk, Northamptonshire, Warwickshire, Worcestershire, Cheshire and Yorkshire. It does not appear to occur in Scotland; but it is reported as abundant in several parts of Ireland, chiefly in the east. It is of considerable interest that in December 1965 a Leisler's bat was disturbed in its hibernaculum in a cavity in a bough during pruning operations in Kew Gardens. This was the first of this species seen in the Gardens despite the recording and observation of over sixty years.

The total range of Leisler's bat is from Ireland through temperate Europe and Asia to China.

Dentition is: $i \frac{2}{3}$, $c \frac{1}{1}$, $p \frac{2}{2}$, $m \frac{3}{3} = 34$.

The Noctule or Great Bat

Family VESPERTILIONIDAE   *Nyctalus noctula*

Though similar to the serotine in size and to the pipistrelle in form, the noctule or great bat was recognised as a distinct species long ago. We might fittingly call this White's bat, for it was the Selborne naturalist who first called attention to it as a native species, under the name *altivolans*, suggested by its high flight. Schreber, however, had some years previously named it *noctula*, basing his description upon a French specimen.

The general form of the noctule is robust and heavy, the fore-arm massive, the wing long and slender, its narrowness being due to the shortness of the fifth finger. The lower leg is short and thick and the foot broad and powerful. The muzzle is broad and has a glandular swelling between the eye and nostril. The nostrils project forward and outward and there is a distinct concavity between the two crescent-shaped orifices. The ear is short—when flattened it is broader than long—with the front border rounded to the tip; its inner surface is covered with short hairs. The ears are spaced widely apart. There is a very short, downy, bow-shaped tragus, broader above than below. The long, soft, golden-brown fur is thick, and extends over the face and over a small part of the wing; it is paler and duller on the lower parts. On the underside there is a narrow band of fur below the arm bones. The last joint of the tail is free. The membrane and ears are blackish.

The head and body together measure up to 3¼ in. (82 mm.) in length. The fore-arm is 2 in. (50 mm.) long. The wing-span is 14–15 in. (350–380 mm.). The weight is unusually variable, from ½–1¼ oz. (16–39 gm.), and the female is larger than the male.

The noctule, as one would expect from the shape of the wings, has a quick, dashing flight recalling that of the swifts, with which, indeed, it may sometimes be seen high in the air, 15–80 ft. (4·5–24·5 m.) up, hawking the same prey just before sunset. It often glides down steeply on expanded wings. It flies at twilight for about an hour and again at dawn as well as in the day-time occasionally. It has a shrill, clear, cricket-like voice when hunting and a strident screech when excited.

Highly gregarious, noctules may number up to 300 in a colony. Some colonies change their roosts at frequent but irregular intervals, especially in June and July, with the result that noctules may sometimes be seen making long flights, perhaps staying in a locality for a few days only before passing on. The longest migration, observed in Germany, was 465 miles (750 km.).

Their roosts are in hollow trees more especially, but may be under the eaves of buildings, where anything up to one or two hundred individuals may associate together, especially during hibernation, which lasts from October to the middle or end of March. Their presence in the roosts is often indicated by thick layers of excrement. Individual noctules may be seen in flight all through the year with the exception of the latter part of December and January.

C. B. Moffat wrote that they 'cram themselves to bursting point either once or twice in the twenty-four hours, during a seventy minutes' career of mad excitement among the twilight-flying beetles and gnats.' They also take moths and other insects, but in captivity they have absolutely refused to eat such 'warningly coloured' species as the cinnabar and magpie moths. It has been shown that at one meal they will consume food equal to ¼ of their own weight. When one considers the lightness of insects, the impact of these purely insectivorous animals on the insect population must be considerable.

The sexes are said to separate into distinct colonies in the early summer, the females retiring into hollow trees to bear and rear their young. Mating takes place from August to October but there is a pause in the development of the embryo during hibernation, and the single young born naked and blind is not delivered until the end of June. Within a month it is independent of the mother.

Although the noctule is generally distributed as far north as Yorkshire, Durham and the Lake District, it is common only in the south of England, from Norfolk to Cornwall, but is rare on the Isle of Wight. It is not recorded from Ireland. Formerly, it was not considered a native of Scotland, but in recent years several specimens have been captured there, as far north as Morayshire. It is found throughout the greater part of Europe and across Asia to Japan and south-eastwards to Burma and Malaya.

The dentition is: $i\ \frac{2}{3},\quad c\ \frac{1}{1},\quad p\ \frac{2}{2},\quad m\ \frac{3}{3}=34.$

The Pipistrelle or Common Bat

Family VESPERTILIONIDAE   *Pipistrellus pipistrellus*

The pipistrelle or common bat is in a general sense familiar to everybody, for it may be seen in the evenings flying everywhere, even in the streets of crowded cities. Its distribution in Britain extends from the south of England to Scotland and the Hebrides, and westwards to Ireland. Its wider range includes Europe and much of Asia. It is the smallest of the British bats.

In spite of its size, the head and body measuring little more than $1\frac{1}{2}$ in. (38 mm.), with just over 2 in. (50 mm.) as the maximum, the pipistrelle is of robust build, and it has a wing-span of 8–9 in. (200–230 mm.). Its fore-arm measures $1\frac{1}{4}$ in. (35 mm.) and it weighs up to $\frac{1}{4}$ oz. (7·5 gm.). The female is slightly larger than the male. It has a flat broad head with a blunt muzzle and a wide mouth. The short broad ears, somewhat triangular with blunt tips and with outer edges slightly notched, end just behind the angle of the mouth. The tail is little over

1 in. (25 mm.) in length, and the legs also are short. The last joint of the tail is free from the membrane and prehensile, and the bat is said to make use of it as a support in crawling up or down. The spur reaches more than half-way to the tail, and an important identification feature is the lobe of membrane behind the spur, known as the post-calcarial lobe. The narrow wing is attached to the middle of the sole of the foot.

The somewhat silky fur, varying from dark to light reddish-brown on the upper parts, slightly paler beneath, extends on to both surfaces of the wings and on to the upper surface of the interfemoral membrane. The wing-membrane and the ears are blackish.

It is a very active bat, flying with a rapid, jerky flight, at heights between 6 ft. and 40 ft. (1·8–12·2 m.) from the ground, on a regular beat over farmyards and gardens and about houses, or up and down lanes, frequently uttering its shrill squeak as it snaps at the flies, particularly gnats, and small beetles, the larger insects being pouched and eaten without alighting. It was formerly supposed that it continued its flight all through the night, but the likelihood is that its activity is intermittent, with periods of rest away from the roost. It has a longer period of activity over the year than any other species, for it leaves its hibernaculum in March and does not retire until the end of October or November. Even then, a moderately high midday temperature or a mild night, with temperature 40°F (4·5°C) or more, is sufficient to awaken it and bring it out for an hour's hunt.

It has a wide choice of sleeping places and is found, singly or in small or large colonies, under roofs, behind drainpipes and gutters, or in crevices between woodwork and brickwork in buildings, as well as in hollow trees and crevices in rocks.

Mating takes place prior to hibernation and fertilisation is delayed until the following spring. The gestation period is forty-four days and one young is born, usually in late June to mid-July, sometimes in August.

It is found everywhere in the British Isles, including the Orkneys and Shetlands. It extends across Europe and Asia to China.

The dental formula of the common bat is: i $\frac{2}{3}$, c $\frac{1}{1}$, p $\frac{2}{2}$, m $\frac{3}{3}$ = 34.

The Barbastelle

Family VESPERTILIONIDAE   *Barbastella barbastellus*

The barbastelle is a medium-sized bat of slender form with relatively long legs and small feet. The ears are large, being as broad as they are long. The inner edges are joined above the eyes, and they are united by their bases between the eye and the mouth, just behind the muzzle. It gives the impression, therefore, that the eyes lie within the ears. The lance-shaped tragus is half the length of the ear, and has two notches towards the base of the hind margin.

The long, soft fur is a very dark brown, almost black, but many of the hairs on the upper surface have pale tips which give a frosted or grizzled effect. This continues on to both surfaces of the wings and the interfemoral membrane. The wing, ear, nose and foot are dusky, appearing lighter than the furred regions, and the under surface is slightly lighter than the upper.

The head and body measure about 2 in. (50 mm.), and the tail $1\frac{3}{4}$ in. (44 mm.). The fore-arm measures $1\frac{1}{2}$ in. (38 mm.). The weight is around $\frac{1}{4}$ oz. (6–8 gm.). This and the long-eared bat are the only British species in which the ears are connected, and the form of the ear in each is so distinct that there is no danger of confusing them.

The barbastelle is usually solitary, but will sometimes associate in groups of half a dozen. It is silent in flight, which is slow and flapping, usually not far from the ground, but may be up to 15 ft. (4.5 m.). This flight begins early in the evening, often before sunset, is continuous until midnight and then intermittent until just before dawn. While in the air the feet appear to be held far apart and the tail decurved. In fine weather it flies high. During its diurnal rest it has been found in various retreats: under thatch of sheds, between rafters and tiles of outhouses, in the crevices of walls and trees, often behind loose bark, and in caves. Its voice is a metallic squeak or a deep buzz.

Hibernation begins in September and lasts until early April.

Nothing is known of its breeding.

The barbastelle is found widely distributed, but nowhere numerous, in all counties south of the Severn and the Wash. It has also been recorded in Cheshire, Warwickshire, Lincolnshire and Cumberland (at

(*right*)
Harvest
mouse with
ear of corn

PLATE 13

(*left*) Wood
mouse or
long-tailed
fieldmouse

(*above*) Noctule
bats at roost

PLATE 14

(*right*) Long-eared
bat at roost

Carlisle), although these records are few. It appears absent from Scotland, Ireland and the Isle of Man. Outside Britain it is found in Europe, from southern Scandinavia southwards, and in temperate Asia.

On each side it has one pre-molar less than the long-eared bat, so that its dental formula stands thus: i $\frac{2}{3}$, c $\frac{1}{1}$, p $\frac{2}{2}$, m $\frac{3}{3}$ = 34.

The Long-eared Bat

Family VESPERTILIONIDAE   *Plecotus auritus*

The long-eared is probably the best known of our bats owing to the very distinctive character afforded by the huge ears, which are as long as the fore-arm and longer than the body. In addition, it is one of the commonest and most widely distributed bats, likely to be met with anywhere in the British Isles, although it is somewhat rarer in the highlands of Scotland than elsewhere.

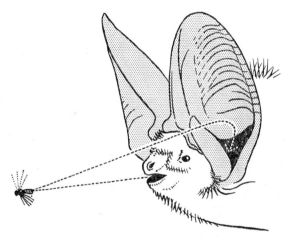

Long-eared bat using echo-location. Dotted line indicates course of ultrasonic pulses

The large and mobile ears give the bat an appearance of size not justified by its small and delicate build. The head and body combined measure up to 2 in. (50 mm.) and the tail is only a fraction less. The fore-arm measures 1½ in. (38 mm.). The weight ranges from ⅙–¼ oz.

6

(5–9·5 gm.). The long ears have their bases joined across the top of the head, and their outer margins end just behind the angle of the mouth. Each is a long oval with rounded tip, and, except for fringes on the folds, they are hairless. They are semi-transparent and are marked with conspicuous transverse folds. The tapering tragus is nearly half as long as the ear, and might be mistaken for it when the bat hangs asleep, for then the ears are folded back under the arms while the tragus stands out beyond the inanimate-looking bundle. Sometimes, when the bat is awake, one ear is held at a different angle from the other, but in flight both ears are directed forward. Sometimes, when the bat is partly awake, the ears are curled sideways and then appear similar to rams' horns.

The silky brown fur is long and thick, especially on the shoulders, but does not exist on the surface of the wings. On the underparts it pales to light brown or dirty white. The wings are both long and broad, and their span is about 10 in. (250 mm.). The long tail, when folded forwards, can touch the top of the head; its tip is slightly free from the interfemoral membrane, and when the bat hangs downwards for sleep it serves as an extra hook for suspension.

The long-eared bat is found chiefly among trees, at heights of 8–40 ft. (2·4–12·2 m.), though it sometimes comes into open windows at night. Its habit is to fly among the branches of trees where it picks insects off the foliage. Its flight is then a glide punctuated with hovering, the body held at an angle of 45 degrees, in order to catch an insect. In passing from one tree to another it flies swiftly and strongly close to the ground. Often, when it has caught an insect, the long-eared bat will land on the ground to eat it. In early spring, when the sallows are in bloom and attracting swarms of insects, the long-eared bat is likely to be there also. The evening flight starts about half an hour after sunset and intermittent flights are made throughout the night, until about an hour before sunrise. Occasionally the long-eared bat may be seen in the day-time. It appears to be at least partially migratory, and a recently published account tells of a group of long-eared bats, coming from the direction of Scandinavia, that settled on a ship in the North Sea. At dusk they took off again heading for the English coast. Moreover, it has been noticed that in summer a swarm will appear in a district where they are not usually seen, and after staying a few weeks will disappear again.

The single young is born in June or July. The breeding females will often associate in large numbers, away from the males, while they are nursing.

Hibernation, which lasts from the middle of October until early April is not so complete as with most other bats. Should the thermometer register 46°F (8°C) or more at any time during the winter, the long-eared bat awakes and makes a foraging flight. On the other hand, it may move its quarters within the cave without necessarily coming into the open. Clusters of hibernating bats are often found in caves or even under roofs of houses, and solitary hibernants may be found in similar situations as well as in hollow trees.

The dental formula is: $i \frac{2}{3}, \quad c \frac{1}{1}, \quad p \frac{2}{3}, \quad m \frac{3}{3} = 36.$

Grey Long-eared Bat

Family VESPERTILIONIDAE   *Plecotus austriacus*

The presence of this European bat in England was first recognised in 1964, from a specimen deposited in the British Museum collection in 1875. This bat had been discovered at Netley, Hampshire. Another was caught at Christchurch in 1909. Since 1964 there have been twenty-four further captures, all in Dorset. This species and the common long-eared bat are so alike that there is some lingering doubt whether the two are specifically different. The grey is said to be more aggressive, biting when handled, in contrast to the docile temperament of the common. The face of the grey is dark to very dark brown, almost black, that of the common being flesh-coloured to light brown.

GNAWING ANIMALS

The Rabbit

Family LEPORIDAE   *Oryctolagus cuniculus*

It is believed that originally the rabbit was a native of south-western Europe, and has spread northwards largely by human agency. It is known to have been introduced to Italy from Spain by the Romans, who are often believed to have brought it to Britain. It is fairly certain, however, that it was the Normans who introduced it here. There is documentary evidence that it was in Britain in the twelfth century, and was then regarded as a desirable domesticated animal, so it was farmed in warrens and zealously guarded. The name 'rabbit' appears to be of Northern French origin, and originally indicated the suckling young, the adults being known as conies.

The ears of the rabbit are remarkably long. If laid forward over the face they nearly reach to the tip of the nose. The eyes are large and prominent and placed well to the sides of the head. The hind-legs are longer than the fore-legs, and strongly developed. In them is the main propelling force in running. Instead of having pads on the soles protecting the feet, all the Leporidae have the feet covered beneath with a thick coating of hair which gives a firm grip either on hard rock or slippery snow. The tail is very short and turned up. The fur is of triple formation. There is a dense, soft, woolly underfur, through which project the longer and stronger hairs which give the coat its colour. There are also still longer but more sparse hairs scattered among the others. The two longer sorts of hair are more or less ringed. The coat becomes thicker in winter.

Black rabbits and, more rarely, other colour mutants, turn up occasionally in wild populations. In a closed community, as on islands or on the mainland where areas are surrounded by barriers to free movement, the genes for colour variation, swamped in an open community, have more chance to emerge.

Rabbits are sexually mature at an early age, and often begin to breed

before they have attained full size. The females are distinguished by
the form of the head, which is longer and more delicately modelled than
that of the male. Bucks are slightly larger than does and may be up to
16 in. (407 mm.) long, head and body, and weigh up to 4½ lb.

Rabbits are not promiscuous
breeders, as has been thought
for so long, but are polygamous,
one buck mating with several
does, each doe keeping to her
own territory within the warren.

Litters of rabbits succeed one
another rapidly at intervals of
about a month from January to
June, but there is some sporadic
breeding in other months of the
year. Ovulation is stimulated by
mating, which is preceded by a
courtship in which the buck

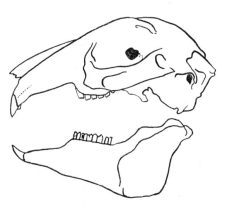

Skull of rabbit with lower jaw

chases the doe. The litters vary from two or three to eight, the higher
numbers being those of the warmer months. Newly born rabbits are
blind and deaf, and almost naked. The ears are closed and have no
power of movement until about the tenth day. The eyes open a day
later. In a few days more the young rabbits can run, and make short excursions from the underground nest. Before they are a month old they
are capable of an independent existence. Until then the mother will
defend them from all dangers, including the weasel and the stoat, using
her powerful hind-feet against her adversary. The weight at birth is
1–1¼ oz. (30–40 gm.). This increases to 9½ oz. (300 gm.) in three to
six weeks.

The gestation period is twenty-eight days. The number of young
born does not correspond with the number of eggs fertilised. In recent
years, it has been established that a percentage of the embryos are
resorbed, which means there is a degeneration of the embryonic tissues,
the substance of which is taken back into the mother's body through
the wall of the uterus. Sometimes whole litters are resorbed. Another
piece of recent information is that thirty-six does, under close observa-

tion, are known to have weaned 280 young, of which 252 disappeared during the season, dead, presumably, from natural causes.

Although so famous as a digger of extensive underground dwellings, some going down as much as 9 ft. (2·7 m.), the rabbit is not specially built for burrowing. Where the soil is light the combined efforts of many generations have resulted in a system of burrows both extensive and complicated, with bolt-runs as emergency exits and stop-runs for nursery use. Although rabbits use the light sand of the dunes, or a sandy heath overgrown with furze and heather, they will also drive tunnels into a firm loam or dry clay. They have also been known to burrow deeply into a surface of coal. The fore paws are principally used in burrowing, the loosened earth being thrown back by the kicking of the hind feet. If stones that cannot be loosened by the paws are encountered, the rabbit has been known to remove these by the teeth. Typical tunnels are about 6 in. (150 mm.) in diameter, increased at points along their length to 1 ft. (300 mm.) to provide passing places. The residential quarters are always blind chambers leading from the main passages. Adult rabbits use no bedding materials but rest on the bare soil. The pregnant doe, on the other hand, makes a bed for her young of hay or straw lined with fur stripped from her underparts.

The instinct to dig seems to be present from an early age and to survive domestication. Baby rabbits being hand-reared with milk from an eye-dropper will sometimes move the paws in the typical digging action while feeding. In addition, domesticated rabbits, after several years of captivity during which time no attempt to dig was shown, have been known to burrow almost as soon as they escape from the hutch. On the other hand, there have long been reports from this country and from Australia, of rabbits' nests, with young, on the surface. Surface nesting became pronounced soon after myxomatosis struck the rabbit here, and one view put forward was that non-burrowing individuals may have been less prone to the disease and by their multiplication a population was produced with a high ratio of surface nesters.

It is usually assumed that wild rabbits are promiscuous, but the detailed account of their courtship given by H. N. Southern indicates a limited polygamy. His observations were incidental to a study made on the productivity and mortality in a population of rabbits within an enclosed warren. The sexes were marked so that they could be recognised with the aid of a telescope, and the markings were made in such a way that he could distinguish and keep records of the behaviour of individual rabbits. At a month to six weeks of age, young rabbits are given to nuzzling and licking each other's heads, but this is not necessarily between opposite sexes. Southern thought he detected 'spheres of influence' for the old bucks, embracing the territories of several does. The buck, apparently, drove all young bucks from his 'sphere of influence'. The does, also, each had an area, which included the entrance to a burrow, and they were found to be highly aggressive towards young does wandering into these territories.

The courtship consisted of half-hearted chasing, with the buck ten to twenty yards behind the doe. The direction taken seemed to be aimless, the two animals loping along, halting frequently but always on the alert. Sometimes the buck would start off with a rush at the doe, who would spurt to maintain her lead, after which both would slow down. At times both would break off to feed, or there would be periods of false feeding (displacement or symbolic activity), when the buck would edge towards the doe until near enough to make a rush at her. It is not obvious whether this is a displacement activity, a symbolic activity or just plain cunning!

These chases were frequently accompanied by other definitive forms of behaviour. The commonest was a tail-flagging when the buck advanced stiff-legged, with haunches up and tail laid across the back. He then moved away from the doe for several yards, showing the white of his tail to her, then returned to her; the performance was repeated several times. Other modifications of this were seen, such as parading in front of her or round her, stiff-legged, with the hind-quarters turned towards her to keep the tail in her view.

Another form of behaviour, known as epuresis or enurination, was that in which the buck would run past the doe but, as he drew level, he would twist his hind-quarters towards her, emitting a jet of urine. At other times he would jump over her, with the same purpose. In some cases the doe reacted 'quite crazily, bouncing and frisking all round the enclosure'.

Aggressive behaviour seemed to take a similar form to that now familiar in other types of animals. A swift direct charge and the intruder usually fled. If not, there would be a leap, with a downward stroke of the paws, to make the fur fly. Enurination was often combined with these manoeuvres, but this seems to be associated with all forms of excitement, even with fear, as in rabbits caught by hand.

Formalised fighting was occasionally noted between two bucks. Southern reported that on one occasion each leapt into the air to meet the other three feet above the ground. Although they came in contact, they seemed to pass each other in the air and land, each in his opponent's starting position; the attack was then launched again from the reverse direction. On another occasion, although the combat was taking place some distance away, the thud of the impact between their two bodies could clearly be heard. Apparently there was no actual striking with the hind-feet, and the encounter seemed to be of the conventional bloodless kind. One bout continued for two to three minutes but with intervals during which the rabbits indulged in 'false feeding', both then turning simultaneously and rushing at each other once more.

The young are usually born in a 'stab' or 'stop', a short burrow about 2 ft. (61 cm.) long, just under the surface, usually well away from the main burrow, the nest being in the blind end. This is a necessary precaution because of the danger that the young may be killed by the buck. If the nursery is in the main burrow the doe guards it at all times against attack from the buck. She visits the nest to suckle the young once in every twenty-four hours and suckles them then for about three minutes. The entrance to a stop is closed with earth after each visit, and some does will spread dried grass over this to hide it. The young start taking solid food at sixteen days and are weaned at thirty days. They continue to increase in size until nine months old. In the first sixteen days they merely double their weight but increase tenfold in the next fourteen days. Sexual maturity is reached in three to four months.

Rabbits are normally silent, except for occasional low growls and grunts, expressive of anger or pleasure. A doe has also been heard to utter low notes when nursing her young. A rabbit, terror-stricken by imminent attack from a stoat, utters a loud scream. Another sound used more deliberately for communication is a thump on the ground with both hind-feet together. This is an alarm signal, usually given by an old buck, to which all rabbits within earshot respond by dashing towards their burrows. The chief enemies, in addition to man, are the members of the weasel family, owls, buzzards, ravens, crows, black-backed gulls, and a variety of hawks. Badgers dig out the young and foxes also take a large toll, as well as wild and feral cats and feral dogs.

Rabbits use special latrines, and accumulations of droppings appear especially upon the flattened tops of old ant-mounds. Refection, which has been called 'chewing the cud', also is used. A rabbit voids two types of droppings, one kind being eaten again (refection), the other discarded at the latrine.

The rabbit is almost exclusively a vegetarian, its chief food being grass and the tender shoots of furze; but in the vicinity of cultivated land it will devastate the crops and inflict serious loss upon the farmer. The exception to a vegetable diet is found in its occasional indulgence in snails and earthworms. Whenever there is sufficient food, and its enemies are not too oppressive, the rabbit will increase its range to the most out-of-the-way corners of these islands. A century ago it was a scarce beast in Scotland, but it was found later in abundance up to the extreme north. It was also found all over Ireland.

In 1954 and 1955, an introduced disease, known as myxomatosis, made the wild rabbit a rare animal. The disease has now become attenuated and there are signs of the rabbit recovering, but usually, as soon as numbers begin to build up in any locality, the disease seems to take its toll once again.

The effects of rabbits' feeding on the countryside were highlighted by the changes following the first wave of myxomatosis. The first results noticed, following the drastic reduction in numbers of rabbits, was that in a year or two green lanes became choked with long grass, seedling trees and brambles, while the margins of cultivated fields were not being eaten bare. On downlands not being cropped by sheep the grass grew taller. At the same time many wild flowers became more plentiful, and this was especially true of orchids.

A report published in 1960 told what had happened in the next stage. Unrestricted growth of the grass led to turf becoming less dense and there was a deterioration in the wild flowers that had so suddenly burst into prominence as the numbers of rabbits fell. There was an increase in brambles, gorse and heather. The most surprising feature of the report was that the vegetation and scenery had reverted to its condition before 1840. It seems that until then there were relatively few rabbits outside the warrens where they had been preserved since Norman times in a state of semi-domestication.

The warren was enclosed by a wall or a fence, except on sand dunes, where it was open. Rabbit fences must allow for the ability of these animals to climb or jump; a standing vertical jump of four feet has been recorded. Mounds were often thrown up for the rabbits to burrow. Strictly speaking, therefore, 'warren' should be reserved for specially stocked systems of tunnels and 'bury' should be used for burrowings away from warrens.

Rabbits can sometimes be seen rubbing their chins on posts and other solid objects, as if they have an irritation. This was investigated by Australian zoologists who found glands under the skin that produced a secretion. The glands are larger in the buck than the doe, and the conclusion reached was that the chin-rubbing is a method of marking territory. Rabbits will also rub their chins on each other, especially on a mate or on the young, presumably to provide a ready means of recognition.

The rabbit used to be placed in a section of rodents known as Duplicidentata, because in the upper jaw there are always two pairs of incisors, there being two smaller teeth behind the two main incisors. All the other rodents have only one pair, and they formed the division Simplicidentata. In 1945, the order Lagomorpha, proposed in 1855 but not generally adopted, was revived to include rabbits, hares and pikas (mouse hares), and to separate these from rodents proper.

The dentition of the rabbits and hares is therefore as follows:

$$i \frac{2}{1}, \quad c \frac{0}{0}, \quad p \frac{3}{2}, \quad m \frac{3}{3} = 28.$$

The Brown Hare

Family LEPORIDAE   *Lepus europaeus*

Although similar in general form and structure to the closely related rabbit, the hare differs in having a longer body, longer hind-limbs, longer ears with their invariable black tips, and tawny fur on the upper parts. In addition, while the rabbit is gregarious, the hare is usually solitary. The two used to be placed in the same genus, but are now considered to be sufficiently unlike to be placed in separate genera.

Skull of hare

The total length of the brown hare is about 24 in. (61 cm.). The weight averages about 8 lb. (3·6 kg.). The shoulders, neck and flanks are of a ruddier hue than the back, which is a mixture of grey and brown. The underside is pure white, except at the breast and loins where the ruddy tint is continued from above. There is a profusion of black and white whiskers, of which the white are the longer. The tail, which is carried curved up over the back or straight behind, is black above and white on the sides and below. The large, prominent eyes have a horizontal pupil and their situation well to the sides of

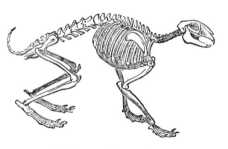

Skeleton of brown hare

the head is assumed to give a wide field of vision. The male, or Jack, has a smaller body, shorter head and redder shoulders than the doe.

The coat is moulted twice a year. The spring moult, which begins in February and lasts until June or July, is protracted and more gradual than the clear-cut autumn moult, which starts soon after the spring moult has ended. There is no great difference between the winter and summer coats, although the former may be rather more grey, the latter being somewhat reddish but with grey patches, especially on the hind-quarters. Both moults start along the back, go down the flanks and limbs and finish at the head and the tail.

The hare does not burrow, and does not seek refuge from its enemies underground unless hard pressed, when it may enter a rabbit burrow temporarily. Instead it relies for security upon speed, or upon crouching in the low vegetation, where its russet coat is not conspicuous and tends to harmonise generally with its surroundings.

The resting place is a slight depression in the long grass known as a 'form', where the hare habitually crouches during the day, watchful and alert, ready to use its superior speed to escape if disturbed. From this it emerges at dusk and goes out to feed, returning to the form at dawn. The feeding time lasts from 19.00 to 07.00 hours but is most intense between 19.00 hours and midnight. Although hares may travel many miles (one was tracked for 30 miles [48 km.] in one night) to feed they usually keep to a home range of one or two miles (1·5–3 km.) radius. The home range is crossed by well-worn trails or pads which can be seen leading to gaps in hedges, walls and fences. A characteristic of behaviour on leaving the form, which tends to break the continuity of the scent-trail, and one used again on returning to it, is suddenly to turn at right angles to the former course and make a prodigious leap. This may be 15 ft. (4·5 m.) or more, to the top of a bank, say, then it will take another long bound, perhaps into marshy ground where the scent will not lie, and from there to the feeding ground proper. Another tactic frequently employed is that of doubling on its tracks. A hare will, on occasion, take to water, showing itself to be a good swimmer, crossing rivers and streams in order to reach a better feeding ground, to avoid pursuit or to seek a mate. A hare has been known to cross a river 200 yd. (183 m.) wide to reach a field of carrots on the other side.

The usual situation for the form is in rank grass among thickets of gorse and briar, or in the open field where the ground is dry beneath it. It takes and retains the shape of the animal's body, and the same form may be used over a period. The doe makes no special nursery, but brings forth her litter of two, three or four young, occasionally more, in the same or a similar form to the one normally used. The leverets are born with their eyes open, and a short furry coat, which lacks the ruddiness of the adult. They are capable of using their limbs within a very short time of birth, and each is sufficiently well advanced at birth to be able to occupy its own form in the vicinity of that of the mother. They are quite independent at a month old. Whether the young wander away to make their own forms entirely unaided, or whether the doe assists by carrying them is uncertain. The former is probable. At all events, the doe visits each in turn to suckle them, and uses the same bounding tactics in doing so, leaving thereby no tell-tale scent-trail.

The hare is exclusively vegetarian, its diet including bark, grain, and roots as well as herbaceous vegetation. Like the rabbit, it can be destructive to young trees in plantations, and where hares are numerous the farmer and the market gardener may lose heavily from their raids on the crops of carrots, lettuce, turnips and other vegetables. In the open country it feeds upon grasses, clover, sow-thistle and chicory, and a hare in a private garden seems to show preference for dahlias, carnations, pinks, nasturtiums, parsley and thyme. In shrubberies it is very destructive to bark and boughs, especially of coniferous trees. As does the rabbit, it occasionally eats flesh, for example, voles.

Refection, the eating of their own droppings, noted first in rabbits, is a regular habit of the brown hare. This takes place during the day, while in the form, but the droppings refected are different from those discarded, and the process may be broadly compared with chewing the cud.

The proverbial expression 'Mad as a March Hare' has reference to the seemingly insane antics of the Jack hares during the rutting season. In spring they lose much of their customary caution and assemble in groups, the males fighting and chasing the females. This may continue in approximately the same spot for several days and the hares can be watched at relatively close range, being apparently oblivious of one's presence. The antics of the males include bucking on stiff legs, kicking, grunting and stand-up boxing matches carried out in pairs, or in groups, at top speed. In bucking one may leap over his opponent and kick him vigorously with the hind-feet, the result being sometimes fatal. If wounded or even only badly scared a hare will scream with a sound somewhat like a child in pain. Another characteristic noise, a warning sound, is made by grinding the teeth. This appears to be passed on from hare to hare and apparently serves the same purpose as the stamping of the hind-feet by the rabbit. The courtship notes of buck and doe are different and their imitation by poachers and gamekeepers is known as hare-sucking.

Hares appear to be capable of breeding all the year round, although there is a peak in spring and summer. In a year there are several litters, usually of four young, but the number varies with the individual and the season. A doe killed in December was found to be advanced with young. The gestation period is forty-two to forty-four days. As in the rabbit there is a heavy mortality among the embryos, which are resorbed, especially in autumn. The young reach sexual maturity at eight months.

There is little information about the brown hare's enemies. The adults may be taken by foxes and in Scotland by wild cats and eagles. All three, however, prey more on leverets, as do stoats and the medium-sized birds of prey. There has, however, long been a belief that hares and rabbits are unlikely to be found in abundance on the same ground. Woodruffe-Peacock in his *Cultivation of the Common Hare* (1905) tells us that when rabbits are abundant they bully, chase and worry hares to death. An earlier writer has told of seeing a particularly vicious attack by a rabbit which sent the hare screaming, bitten severely on its hindquarters. Harry Thompson in *The Rabbit* (1956) suggests that this may be true. Colin Matheson gives figures for a Glamorgan estate where rabbits increased from 6,353 in 1879-83 to 17,006 in 1909-13, while hares diminished from 738 to 22 in the same period.

Contrasting with the apparent submission to a rabbit's aggression, the hare has been reported to show fight unexpectedly. A doe was once seen, when a cow inadvertently pushed her muzzle towards the leverets, to rise on her haunches and box the cow with her paws. There was another occasion when a man found himself attacked three times by a hare; he discovered later that he was standing near her leverets. And a rook that ventured too near a leveret was said to have been killed by the doe.

The brown hare is widely distributed in England, Wales and Scotland, up to about 2,000 ft. (600 m.) elevation, except in Scotland where the blue hare, *L. timidus*, replaces it from 1,000 ft. (300 m.) upwards. It is not a native of Ireland, which has instead a subspecies of *L. timidus*, known as the Irish hare.

Dental formula: i $\frac{2}{1}$, c $\frac{0}{0}$, p $\frac{3}{2}$, m $\frac{3}{3}$ = 28.

The Scottish Hare

Family LEPORIDAE   *Lepus timidus scoticus*

The Scottish hare is regarded as a subspecies of *L. timidus* of Scandinavia and the Alps. Known also as the blue or variable hare, it is indigenous only in Scotland and has been introduced into some of the neighbouring isles, and also into northern England, north Wales and Northern Ireland.

The name variable hare denotes its change of hue at the beginning of winter after the manner of the stoat. In Cheshire it is known as the white hare. The winter whitening of the fur was formerly thought to be like the whitening of human hair in old age, due to the activity of certain mobile cells, which removed the pigment. The probability is that a process takes place similar to that seen in stoat and weasel, which has been more fully investigated (see pp. 141–2). It is noteworthy that the black tips of the ears, like the black tip of the tail of the stoat, never change colour. Moreover, the outer half of the ear below the black tip remains brown, and there is always an area of grey, of variable size, on the back. The Scottish hare is smaller than the brown hare, the combined length of head and body being about 20 in. (51 cm.), but the head is proportionately larger, the ears and tail shorter, and the legs longer. The weight is 7–8 lb. (about 3·5 kg.) in fully grown individuals. The fur is more woolly and of a greyer tint in summer, the whiskers shorter and finer, the eyes rounder, and the hair on the underside of the foot softer. Behind the breast the underparts are white, and the tail is completely so.

The habits of the Scottish hare are very similar to those of the brown hare, except that it lives mainly above the line of cultivation. In general its food is similar, but it is said to feed on lichens in winter. Instead of making a form, it hides in rock crevices and among stones where it may be sheltered from the sight of birds of prey overhead. The breeding habits do not appear to differ greatly from those of the brown hare. Breeding begins in February, reaches a peak in March and falls off in July. Gestation is about fifty days. Two to three litters are produced in a year, and the leverets vary in number up to eight.

Dental formula: i $\frac{2}{1}$, c $\frac{0}{0}$, p $\frac{3}{2}$, m $\frac{3}{3}$ = 28.

The Irish Hare

Family LEPORIDAE   *Lepus timidus hibernicus*

The Irish hare occurs, naturally, all over Ireland, where it is common in the mountainous parts, although it is sometimes found on lower ground. It is not found elsewhere except in north Wales and the Island of Mull, where distinct attempts have been made to introduce it.

Although it is larger than the Scottish hare, there is a tendency today to identify the Irish hare completely with the Scottish hare and to ignore the subspecific names. The head and body average about 23 in. (58 cm.) in length, and the tail about 3 in. (76 mm.). The ears slightly exceed the tail in length. The average weight is about 7 lb. (3·2 kg.). It has russet fur, not smoky brown as in the Scottish hare. The winter whitening is less regular than in that subspecies, the winter coat differing little from the summer except in the greyer rump. Any greater change than that is patchy, with russet 'islands' being left surrounded by white. As compared with the brown hare, the Irish hare is smaller and of more graceful build, but the head is relatively longer and broader, the eyes rounder, the ears shorter and the limbs longer.

Though it does not dig burrows of its own, it has been known frequently, when coursed, to take refuge in a rabbit's burrow. Like the other hares, the Irish hare is solitary, but it shows a tendency to gregariousness at times, especially in winter and in the breeding season, and, on occasion, has been seen in the north of Ireland moving in droves of two or three hundred.

It has several litters during the year, averaging three leverets a litter.

(*above*) Brown hare in marshy habitat in early spring

PLATE 15

(*above*) Brown hare
leverets eight days old

(*right*) Brown hare

(*left*)
Ship rat
(commonly
called black
rat) in
granary

PLATE 16

(*below*)
Common rat,
also known
as brown rat

They seldom remain together long, either moving apart of their own accord or being separated by the old doe. They are able to run when only an hour or two old.

Dental formula: i $\frac{2}{1}$, c $\frac{0}{0}$, p $\frac{3}{2}$, m $\frac{3}{3}$ = 28.

The Dormouse

Family GLIRIDAE   *Muscardinus avellanarius*

The dormouse looks like a miniature squirrel in the shape of its head, its prominent black eyes, large ears and thickly furred tail, as well as its arboreal habitat and its habit of sitting up on its haunches and holding a nut or other food in its fore-paws. Superficial resemblances are often due to a similarity of habit and habitat. However, in its anatomy it shows a closer affinity to mice. Its total length is about $5\frac{1}{2}$ in. (139 mm.), of which nearly half is tail. The weight is variable, from $\frac{3}{4}$–$1\frac{1}{3}$ oz. (23–43 gm.), being greatest just before hibernation. The fore-limbs, which are much shorter than the hind-limbs, are furnished with four separate fingers and a rudimentary thumb, while the hind-feet have five

Skeleton of dormouse

toes, of which one is vestigial. All the claws are short and on the underside of each foot there are six large pads. The fur of the upper part is light tawny, that of the underside yellowish-white, but on the throat and adjoining part of the chest it is a purer white. The head is comparatively large, with a blunt muzzle, prominent eyes, broadly rounded short ears and long whiskers. It can be found in the copse, the thick hedgerow, scrub and secondary growth in woods, where there are trees, such as hazel and beech, with edible seeds. The dormouse seems to be almost wholly nocturnal, although it can occasionally be seen active by day. The daily sleep approaches the deep sleep of hibernation,

7

with a heavy fall in the body-temperature and in the rate of breathing, so that the animal gives the appearance of being actually dead. Most animals sleep 'with one eye open', but the sleeping dormouse has to be shaken to be aroused.

Three to five globular sleeping nests are built in a season, each made of twigs, moss and grass, about 3 in. (76 mm.) in diameter, and sometimes with a round opening. Although they may be sited among the stubs of a coppice, or under a tussock of grass, they are typically suspended high up in bushes. The nursery nest is twice that size, in some districts constructed of the bark of old honeysuckle stems, shredded into ribbons. The inner lining is of the same material more finely divided, with, finally, a bed of leaves.

Little is known of its breeding habits except that the gestation period is twenty-two to twenty-four days and that the female drives the male away from the nursery nest, as in so many small rodents, to occupy a sleeping nest on his own. There appear to be several litters a year of three or four, or even six or seven, blind and naked young, born from May onwards, and there are records of young being found in September or October. What happens to the young born so near to the period of hibernation is an open question. It has been suggested that they probably do not survive, simply because they are not able to provide their own shelter or to go without food for a long period. However, in view of the many surprises brought to light in recent years in the matter of hibernation, their survival cannot be ruled out as impossible. The first coat which they acquire by the age of thirteen days, is more grey than red, but at eighteen days they moult, the new coat being like that of the adult, but paler. Their eyes open at eighteen days, and they start to forage. They become independent in forty days or less and sexually mature at about one year.

Hibernation is from late October until April, each individual hibernating on its own. During the period preceding this there is an accumulation of fat in the body, and further provision for the winter is made by laying up a store of nuts. The winter sleep is not continuous; the dormouse wakes at intervals for a meal, and then resumes its sleep. The winter nest is usually under moss at the base of a tree or in a hedge-bottom, commonly under vegetable litter, such as dry leaves. It is also said to hibernate occasionally far underground. The onset of hibernation appears to be uninfluenced by temperature or other atmospheric conditions. The accumulation of fat is, probably, the sole predisposing factor.

In hibernation the dormouse rolls itself into a ball, the chin resting on the belly, the hind-feet brought forward level with the nose, the fore-feet clenched and held either side of the muzzle, the ears folded back on to the head, and the tail brought forward between the hind-legs with the tip wrapped over the face and back. Its temperature drops so low that the body feels cold. The pulse is slow and feeble and the muscles are held rigid. A sleeping dormouse can be rolled across a table without giving any sign of being disturbed.

The food of a dormouse is much the same as that of a squirrel, but it is particularly fond of the hazel-nut, a good fat producer, beech mast and chestnut, pine seeds, haws, young shoots and bark. It does not crack the shell of the nut, but gnaws a small hole in it. In addition it may eat insects, and has been said sometimes to take birds' eggs or even young birds.

The dormouse used to be a favourite pet, largely perhaps because it is inoffensive and shows little disposition to bite, but it seems to be more rare now than formerly, perhaps the result of competition for food with the grey squirrel, or from causes not yet determined. There is a high mortality in winter when it falls prey to magpies and carrion crows, foxes, badgers, stoats, weasels and rats, possibly also grey squirrels. Four out of five dormice, it has been estimated, are killed during hibernation. Its maximum span is probably four years.

In Europe the dormouse is found no farther north than Sweden and it does not occur in Scotland. It is absent from Ireland. In England it is found mainly in the south and west, being absent from East Anglia and infrequent in the Midlands. In Wales it is scarce and somewhat local. Outside Europe it extends only to Asia Minor. The English name can be traced back certainly to the fifteenth century, and it is considered to embody the verb *dorm*—to doze, still used in the north of England, which brings it very close to the sleepmouse of southern England and sleeper of other parts. Derrymouse, dorymouse and dozing-mouse are other local variants.

The dentition is much the same as that of the squirrel. There is a single large incisor on each side of the upper and lower jaws, one pre-molar and three molars after a considerable gap:

$$i \frac{1}{1}, \quad c \frac{0}{0}, \quad p \frac{1}{1}, \quad m \frac{3}{3} = 20.$$

The enamel ridges of these cheek-teeth constitute a rasping surface such as no other mammal possesses.

The Edible, Fat or Squirrel-tailed Dormouse

Family GLIRIDAE   *Glis glis*

The largest of the European dormice, with a total length of 14 in. (356 mm.), the edible dormouse was introduced at Tring, in Hertford-shire, in 1902. Walter Rothschild had brought some from Germany and Switzerland and liberated them in Tring Park. The bushy tail, which is as long as the rest of the body, and the grey pelage cause confusion with the grey squirrel, but the dormouse is smaller and is nocturnal.

Normally inhabitants of deciduous woodland where they live on fruit, nuts, insects and bark, the edible dormice often invade houses where they seek places to hibernate. Prior to hibernation in October weight is increased, but there will be up to 50% loss of body weight by the time they wake in April. Breeding starts in mid-June with one litter of two to seven young, in a nest in a tree-hole.

Until 1963 the range of the edible dormouse was within a triangle bounded by Beaconsfield, Aylesbury and Luton. There was little indi-cation that they did any damage although Public Health Departments were often called in to remove edible dormice that had taken up residence in houses. However, in 1963 the dormice were found to be causing extensive damage on a Forestry Commission estate at Wendover. Damage to conifers, especially spruce, was sometimes causing the tops of the trees to die.

Dental formula: $i \frac{1}{1}, \quad c \frac{0}{0}, \quad p \frac{1}{1}, \quad m \frac{3}{3} = 20.$

The Bank Vole

Family MURIDAE   *Clethrionomys glareolus*

The bank vole frequents hedgerows and wooded country rather than open fields. Its head and body length is $3\frac{1}{2}$–4 in. (88–101 mm.), the tail being nearly half this length and ending in a pencil of hairs. The ears and feet are proportionately large, the former also being more oval than round. Bodily proportions are variable and so is the weight, from about $\frac{1}{2}$ oz. (15 gm.) in winter to double that figure, or more, in summer. There is an important anatomical difference between the bank vole and other voles: the molar teeth become rooted in the jaw in the adults.

The fur of the upper parts is a bright chestnut-red or vandyke brown, except the hairy tail, which is black above. The underparts, including the lower side of the tail, are whitish varying to yellowish or even buff. The redder tint causes this species sometimes to be referred to as the red vole. The lips are pink, the feet grey, and the whiskers are about 1 in. (25 mm.) long. Black and albino varieties have been recorded. Moulting can be seen at all times of the year, but there are peaks in spring and autumn.

The bank vole is much more agile than the field vole, but less given to jumping or burrowing. It may be seen in sunny situations at almost any time of the day, preferring warm dry places, but not infrequently found in wet areas. It is a good swimmer and diver. Its shallow runs are constructed in the earth of a roadside bank or hedgebank. They have many entrances and exits at different levels, some of the passages connecting with the top of the bank, others enlarging into blind chambers. The bank vole's diet includes herbage, roots, bulbs, fruits and seeds; it appears to be particularly fond of turnips. In spring it has been observed climbing rose and hawthorn bushes in order to nibble the new leaves, and in autumn to obtain the hips and haws. It also seeks nuts, berries, grain of wheat and barley, and seeds of smaller grasses. Insects and larvae make up about a third of its food, and snails and even small birds are eaten. The entrance to its burrow frequently gives evidence of the variety of its food. It has been known to eat the unpalatable shrew that it has itself killed, and is said to have given way to cannibalism. These suggestions of cannibalism in mammals must be treated with reserve. There is little to show that it is at all widely spread in any species, rather that it occurs under unusual circumstances. Too often such suggestions are based upon animals in captivity, where apprehension and often over-crowding or shortage of food cause a perversion of natural instincts. The records of cannibalism in the bank vole are, however, from observations on wild individuals. In Scotland it is accused of eating the shoot-buds of young conifers, especially of larch, and of gnawing the bark from branches. In Britain it is occasionally caught in the act of raiding household stores. In more northern latitudes, as in Norway, it is a constant inhabitant of houses. It does not hibernate, and it does not, as a rule, lay up food stores. It is markedly diurnal in thick cover, but probably has alternating periods of activity and sleep throughout each twenty-four hours. It is said to be more active at night in summer.

It is preyed upon by weasels and tawny owls especially, but falls victim to most other predatory birds and mammals.

The breeding season starts at the middle of April, rises to a peak in June, then declines in intensity until October, but there may be a limited amount of breeding in all months between October and April. The males have the appearance of being quarrelsome. When fighting or pairing they are very vocal, indulging in grunting squeaks. The gestation period is about eighteen days. There are four or five litters of three to six, or more, naked and blind young during the year, produced in nests of grass, moss and wool, or feathers, usually placed above ground, sometimes in a bird's nest at a fair height. The young are weaned at two and a half weeks, become sexually mature at four to five weeks and produce a second summer generation which reaches maturity the following spring, to produce the first summer generation. There can, however, be some breeding in winter. Little is known about longevity.

The bank vole appears to be widely distributed, and to occur as far north certainly as Moray and Elgin; but it is not recorded from Ireland, the Isle of Man, the Hebrides or the Shetlands.

Dental formula: i $\frac{1}{1}$, c $\frac{0}{0}$, p $\frac{0}{0}$, m $\frac{3}{3} = 16$.

The Short-tailed Vole

Family MURIDAE   *Microtus agrestis*

This was formerly called the grass mouse, field mouse or field vole*. The general stumpy form of the short-tailed vole, with the blunt oval outline of the head, the short, round ears just protruding from the reddish-brown fur of the upper parts, and especially the short, rather stiff tail, are points sufficient to distinguish it from either of our mice. On the underside the fur is greyish-white. Moulting seems to take place the year round with ill-defined peaks in spring and autumn. The hind-feet have six pads on the under surface, as compared with the five of the water vole. The length of head and body is $3\frac{1}{2}$–$4\frac{1}{2}$ in. (88–114 mm.), the tail being a third of the body length. The weight varies from $\frac{2}{3}$–$1\frac{1}{3}$ oz. (20–40 gm.).

The typical habitat of the short-tailed vole includes meadows and damp pastures, but it will also be found in open woodlands among bracken, in gardens, orchards and plantations. In such places it does enormous damage, its food being almost entirely the hard stem and leaves of grasses, although it will nibble almost anything vegetable, including bark. It is, however, said to consume large numbers of insects among them the destructive larch sawfly. It lays up extensive stores of food in its underground burrows, but it is not correct to say that the underground burrows include the summer nest. The burrows connect with a network of runs above ground through the grass and herbage, with occasional holes enabling the vole to bolt underground. These surface runs are made without disturbing the grass blades, which hang over them, screening the runs from above, so enabling the vole to creep or run along without being seen by a hawk circling high overhead.

It is possible to stand or crouch over a patch of coarse grass for long periods, knowing that voles are present in the grass, and yet not see a sign of them. There will be slight movements of the blades of grass, and every now and then a stem will be pulled down and drawn into the tangle of grass concealing the vole. That is all. The best way of appreciating the position and extent of the runs is to examine ground over

---

*The name 'vole', now in common use, is merely a shortened form of voldmouse and this, ironically, merely means field mouse.

which has been a grass fire. Then, among the blackened stumps of the tussocks can be seen a close and extensive network of the runs.

The short-tailed vole is less successful in eluding the owl, which hunts much nearer to the ground, and probably does so more by hearing than by sight, and the weasel which can follow it into its runs. Other enemies include the kestrel, buzzard and barn owl; foxes, too, even before myxomatosis, would often feed on this one species.

It is often suggested that an owl hoots to frighten ground prey, bringing them into the open where they can be captured. This is probably not strictly true. I have several times had a short-tailed vole under observation at dusk, and seen it freeze the moment an owl hoots; it would probably do the same when under cover.

The female vole makes her nest beside a rank tuft of grass along one of these runs. It is roofed with a dome of grass blades divided longitudinally and plaited and felted. There is little to distinguish it from the surrounding dried grass litter so that only a trained eye would see it; or one may find it accidentally by groping among the litter. Once detected, the fine texture, due to shredding of the grass used in its construction, is readily apparent. The female enters or emerges from any point under the edge of the dome, and if the nest is uncovered suddenly, will at once bolt, leaving her youngsters unprotected. Characteristically, when the nest is disturbed, the most we see of the mother is a shape which flashes away and disappears into the adjacent grass. The nest may sometimes be found by rolling away a large log. In this kind of situation it is usually without a dome, and is evenly cup-shaped, like a bird's nest.

The breeding season lasts from late February to September. Gestation is twenty-one days. Like all our rodents, the young voles are naked and blind at birth. There may be five, six or seven in a litter with several litters in a season, from late February to September. The young are weaned at fourteen to eighteen days, are sexually mature at three weeks, and mate at six weeks of age.

In districts where weasels and owls, as well as other predators, have been more or less exterminated, the natural increases of the short-tailed vole are allowed full play, and they can become a plague. Pastures are eaten bare. Crops are cleared from the fields, young trees in plantations destroyed by the thousand, and even newly sown cornfields rendered

(*above*) Red
squirrel

PLATE 17

(*right*)
Weasel in
typical
attitude of
rearing up
and
reconnoitring

(*above*) Water vole on river bank

PLATE 18

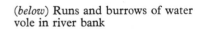

(*left*) Orkney vole

(*below*) Runs and burrows of water vole in river bank

(*above*)
Badger and
cub leaving
set

PLATE 20

(*left*) Vixen
(Note
elliptical
pupils to
eyes)

(*right*) Harvest
mouse washing
forepaws

PLATE 19

(*below*) Harvest
mouse and nest
among wheat

unproductive because every seedling has been eaten. On these occasions, the ground being eaten bare and the voles so numerous, the surface of the soil appears to be on the move. In the New Forest and the Forest of Dean great loss has been sustained at various times by the voles severing the roots of young trees that crossed their runs, and by gnawing the bark of the young trunks. The most effective of the plans adopted for lessening their numbers was to sink pits 1½ ft. (45·5 cm.) deep, wider at the bottom than at the mouth, into which vast numbers fell and from which they could not escape. Towards the end of the last century, the south of Scotland suffered from a plague of so-called 'mice' that ate up everything in the fields, inflicting such serious loss to agriculture that in 1892 a Government Commission of Investigation was appointed to inquire into it, and it was found that the chief culprit was the short-tailed vole. At the peak of such plagues, large numbers of short-eared owls move into the area, as do other predators, such as kestrel, buzzard, barn owl, harriers and fox, and these remain for a year or two after the numbers have waned. It has been suggested that the enormous increase in the numbers of the voles on these occasions is directly due to the warfare waged by keepers on weasels and owls. The vole plagues, however, have been more carefully investigated during the last thirty years. These violent fluctuations in numbers, with the population building up to a peak and then 'crashing' until the normal is again reached, are a feature of many species of rodents. The cycle for the short-tailed vole includes a peak every four years. The causes of the increases are less readily understood than the causes of the cataclysmic decreases. The destruction of predators is only one factor involved, and the influx of predators at the peak is probably the least important factor in the decrease. One of the probable reasons for the sharp decrease may be that when the overcrowding occurs fighting between themselves consequently becomes more intense. This probably interferes with normal breeding; it may bring higher mortality among litters that are produced. And there is the chance that the ground is fouled so that the vegetation is 'sick'.

Incidentally, it was shown by the report of the committee, in 1893, that the rook destroys great numbers of field voles, not only the adults that chance to cross the fields where the rooks are digging for cockchafer grubs, but also the nests and young which the rooks systematically search out.

As with harvest mice and shrews, the short-tailed field vole has alternating periods of feeding, other activities, and sleep, the periods being two to three hours, throughout each twenty-four hours, those of the night being slightly longer than of the day.

The short-tailed vole is absent from Ireland, Isle of Man and the Scillies, as well as most Scottish islands. Several subspecies have been recognised but these are of doubtful value.

Dental formula: $i \frac{1}{1}$, $c \frac{0}{0}$, $p \frac{0}{0}$, $m \frac{3}{3} = 16$.

The Orkney Vole

Family MURIDAE   *Microtus arvalis orcadensis*

So far back as 1805, the Rev George Barry, in his *History of the Orkney Islands*, mentions a rodent that was known locally as the vole mouse. He says it 'is very often found in marshy grounds that are covered with moss and short heath, in which it makes roads or tracks of about 3 in. (76 mm.) in breadth, and sometimes miles in length, much worn by continual treading, and warped into a thousand different directions.'

Towards the end of the last century, J. G. Millais obtained specimens, and on critical examination found that they differed from known forms in several details, sufficient in his opinion to constitute a new species, which he called *M. orcadensis*. Later systematists have regarded it as a subspecies, for it was found subsequently that specimens from various islands in the Orkney group show small differences, although generally speaking they are much alike, and their habits are practically identical, so far as is at present known.

The runs of the Orkney vole, among the heather and the rough vegetation, running along the surface and at intervals entering tunnels about $2\frac{1}{4}$ in. (57 mm.) in diameter, are a conspicuous feature of the islands. Their nesting places are under small mounds connected by a network of runs, or they may be on the surface. The nest itself is made of grass and roots in a rounded chamber, where at intervals during the spring and summer several litters, varying from three to six, are produced. Before they are three weeks old they are capable of independent existence.

The Orkney vole appears to be specially fond of the roots of heath rush, but also feeds on grass and the crops in cultivated fields to which it can gain access. It has many enemies to hold its increase in check, for every bird and beast large enough to capture it will eat it readily, although it is preyed upon more especially by hen harriers, short-eared owls and other predatory birds.

Dental formula: i $\frac{1}{1}$, c $\frac{0}{0}$, p $\frac{0}{0}$, m $\frac{3}{3}$ = 16.

The Water Vole

Family MURIDAE   *Arvicola terrestris*

The water vole, a comparatively inoffensive rodent, is apt to suffer from its folk name 'water rat', yet it would seem that this is the animal with the best claim to the name. 'Rat' is Anglo-Saxon, so it must have been in use before the ship rat or the common rat were known in this country. The only animal to which it could have applied must surely have been the one we are dealing with now: the water vole. It has the characteristics of a vole: a short thick head, with the muzzle rounded instead of pointed; the limbs relatively short; the eyes small and extremely short-sighted; and the small round ears scarcely projecting beyond the surrounding fur. As to the short-sightedness, I have on several occasions brought my face slowly to within a foot of the face of the water vole, looking straight into its eye and the vole would continue feeding. At the slightest sound or sudden movement it would dive under water at once. The long, thick, glossy fur may be of a warm reddish-brown above, sprinkled with grey, but it is commonly a blackish-grey. On the underparts it is yellowish-grey. The feet are naked, pale pink on the underside with five rounded pads and clothed with stiff hairs on the upper surface. The head and body measure 7½–8½ in. (19–21·5 cm.), the tapering and hairy ringed tail is about 4½ in. (114 mm.). Weight varies from 3¾–5¾ oz. (120–180 gm.) in winter to almost double this in summer. Females are slightly smaller than males and more greyish-brown. *A. terrestris brigantium*, with forwardly directed incisors, is a subspecies confined to Yorkshire. There is also a subspecies in Scotland, north of the Clyde (*A. terrestris reta*) which is characterised by its smaller size and its blacker fur. *A. terrestris*, however, is variable throughout its range and too much emphasis should not be placed on these supposed subspecies.

The water vole does not hibernate, but it has been said to lay up considerable stores for when food is scarce and difficult to find. These stores are said to consist of nuts, beech mast, acorns, and the creeping underground stems of horsetails. Although it is aquatic, steady rain will keep it in its burrow or cause it to gather food from near the mouth of the burrow and take it inside, presumably to consume it there and then. Water voles have been reported eating caddis-worms and other insects and they take freshwater snails and mussels. Often collections of the empty shells on a bank show where they have been brought ashore to be eaten. Apart from these, its food is mainly vegetable: succulent grasses, flags, loosestrife and sedges and other plants growing along river margins.

The water vole is usually described as mainly nocturnal. My own observations over a period of weeks in one locality suggested that it has a four-hourly rhythm of activity throughout the day and night, with feeding periods of about half an hour alternating with periods of rest or random movement. Observations by other people tend to differ from this and from each other. The water vole is sometimes found in fields far away from water, and it is noteworthy that the Continental water vole is much more terrestrial than the British subspecies. The home range is marked with a scent from glands on the flanks conveyed to the ground by the hind-feet.

Near a stream or lake the sudden 'plop' as it drops into the water is the observer's first intimation of the presence of a water vole. Occasionally we can track its course under water, but as a rule it at once disappears, and surfaces some distance away or retreats into a burrow in the bank, sometimes by an underwater entrance, or regains the bank by an upper exit. It is a steady swimmer, its rate of progress being an even $2\frac{1}{2}$–3 m.p.h. (4 km.), but it is less skilful in swimming than in diving. However, this 'plop' is such a characteristic sound that, even at night, it is an indication of the water vole's activity; and the frequency with which it is heard at night tends to invalidate one suggestion, that water voles show little activity at night.

The burrows have been said to cause considerable damage to the dykes in Fenland, and where ponds have been constructed by artificial banking. Otherwise, the water vole is largely inoffensive.

The breeding season lasts from early April to October. The gestation period is probably three weeks as in related voles. The female makes a thick-walled globular nest of reeds and grasses in a chamber under the bank, or in a hollow willow or even in a bird's nest, for her litter of about five (two to seven) naked and blind young. The number of litters in a season is not known for certain, but the young of early litters mature quickly and breed before the winter. The life-span is little more than a year in the wild because the older individuals are driven out of their territories by the younger voles and more readily fall prey to herons, owls, otters, stoats and weasels, rats, pike, eels and large trout.

The water vole is generally distributed in Britain, but does not occur in Ireland or the Scottish islands.

The surface of the molar teeth in all the voles presents a pattern of alternating triangular prisms. In the water vole and the field vole these teeth are not rooted in the jaw; in the bank vole they are in the adult.

Dental formula: $i \frac{1}{1}$, $c \frac{0}{0}$, $p \frac{0}{0}$, $m \frac{3}{3} = 16$.

The Harvest Mouse

Family MURIDAE   *Micromys minutus soricinus*

With the exception of the pygmy shrew, the dainty harvest mouse is the smallest of British mammals. The harvest mouse will always be associated with the name of Gilbert White, who first made it known as a British mouse. A description of it was first published by Pennant in his *British Zoology* (1768).

The head and body combined measure 2–3 in. (50–69 mm.), and the nearly naked, scaly tail is almost as long. The weight is $\frac{1}{8}-\frac{1}{4}$ oz. (4–10 gm.). The thick, soft fur of the upper parts is yellowish-red in colour, and of the underparts white, the two colours being fairly sharply separated. The tail is pliant and the outer end is prehensile. The moment a harvest mouse ceases to move, its tail seeks to wrap itself around the nearest support. While the mouse is moving, the tail is all the time taking a partial grip on any support available, rather in the way we use the hand on a handrail, not actually gripping but ready at a moment's notice to do so. It serves, therefore, as an additional foot. In addition the outer of the five toes on each hind-foot, the one corresponding to our little toe, is large and opposable to the rest. A typical attitude for a harvest mouse is to grip two adjacent stems with the hind-feet and wrap the tail around one or both of the stems, leaving the front paws free for holding food. The harvest mouse has bright black eyes, a short blunt nose, and short rounded ears, the latter about one-third the length of the head.

The typical habitats of the harvest mouse are pastures and cornfields, where it climbs the stems of the tall grasses and corn plants, cutting off the ripe ears and carrying them to the ground where it devours the grain. It has also been found in open fields, salt marshes, reed beds and dykes. During the summer it is said to feed to some extent on insects, but it will also eat the seeds of a wide variety of grasses. In this same period it stores grain in burrows for winter use. Sometimes, however, instead of wintering in burrows in the earth, it tunnels into hayricks, and if undisturbed may even bring up a litter or two there. As a rule it constructs the wonderful nursery, which has won human admiration ever since White made the species known.

This is a spherical nest, about 3 in. (76 mm.) in diameter, of woven blades of grass. It has no definite opening, the grass blades being merely pushed aside to make entrance or exit where required, and closing again by their own elasticity. There is just sufficient room inside for the female and her blind and naked offspring. The male is not allowed to enter. The nest is slung at heights of a few inches to a foot or more (60–300 mm.) above the ground, from several stout stems. Alternatively, it may be lodged between the stem and leaf of a thistle, or a knapweed, or in the branches of a blackthorn bush or broom. The bed is made of leaves of corn or grass split longitudinally. Harvest mouse nests are not always so tough as that described by White, which 'was so compact and well filled that it would roll across the table without being decomposed, though it contained eight young.'

Breeding is probably from April to September. The gestation period is twenty-one days. Several litters of five to nine young are produced a year. The eyes open at eight days. Excursions from the nest begin at eleven days and juveniles are independent at fifteen days. The expectation of life is about eighteen months, but two individuals in captivity lived for five years.

It is usually said that harvest mice are more diurnal than other mice, yet four individuals kept by me under natural conditions within ample cages, so that their habits were not upset, showed a three-hourly rhythm. Every third hour, night and day, they fed for about half an hour. At the middle of each three-hourly period they were asleep, under cover, and the sleeping periods occupied from one and a half to two hours of the three-hourly period. It might be broken as one or the other came out to take a run round after which it went back to sleep. It was noticeable that on such occasions the mouse usually took a drink before going back to sleep.

Until about December the young of the year resemble the house mouse in colour, and may easily be mistaken for it. Then from the hind-quarters forward they begin to assume the reddish tint. As the adult harvest mouse weighs only about $\frac{1}{6}$ oz. (6 gm.), it is not surprising that it should be able to sit on an ear of corn to which the grasping feet and prehensile tail have enabled it to climb with ease. It will also run up a very slender grass stem, which gives under it. This does not spill the mouse which rapidly swings its tail from one side to the other, producing a balancing effect like a tightrope walker with his long pole extending either side of him. In spite of its common name, this mouse is not only found in or about cornfields. It also frequents the tall, rank herbage along ditches and untrimmed hedgerows. In winter it used to be most commonly found about the lower parts of wheat and oat ricks.

Generally inoffensive and gentle, the harvest mouse is said at times to become savage and cannibalistic. Having kept several of these mice for five years, I find it hard to believe this. It lacks the offensive odour of the house mouse, which is one point in favour of it as a pet. Its voice is a low chirp.

Harvest mice are said to have decreased greatly in numbers as well as in the area of their distribution during this century. It has been suggested that the use of reaping machines, giving a shorter stubble than the scythe, was responsible. There was, however, a revival in 1955. There are records of the occurrence of harvest mice in nearly all the counties of England and Wales, in the past, and from parts of the Scottish lowlands. Today, they appear to be most abundant in the south of England, becoming more rare as we go north. Harvest mice do not occur in Ireland.

It is, however, worth recalling the early history of this species. It was not made known to science until 1771 when the Berlin-born naturalist, Peter Pallas, who worked for the Russian government, found it on the banks of the Volga. Meanwhile, Gilbert White had discovered it in Hampshire, in 1767, and in the same year George Montagu found it in Wiltshire. Neither of these discoveries was made public until 1789 when White published his *Natural History of Selborne*, although Thomas Pennant had mentioned in his *Zoology* (1768), 'the less long-tailed field-mouse'. This suggests that the mouse may have been tolerably rare even before the eighteenth century, sufficiently rare, that is, for all but the more persistent naturalists to have overlooked it.

Dental formula: $i \frac{1}{1}$, $c \frac{0}{0}$, $p \frac{0}{0}$, $m \frac{3}{3} = 16$.

The House Mouse

Family MURIDAE   *Mus musculus*

The most familiar, the most widely distributed and one of the most numerous of our mammals is the house mouse. Although of a seemingly timid and retiring nature, it can on occasion exhibit a certain boldness. It is readily tamed, and if not molested or disturbed it will largely ignore the presence of human beings with whom it lives for preference, although it is often found in woods and fields. For the most part, however, this rodent is found in the immediate neighbourhood of buildings, especially where there are stored foods, corn ricks being particularly attractive to it. There are even colonies of mice in the large meat refrigerators, in perpetual darkness and at temperatures below freezing. They feed solely on meat. They are larger and heavier, have longer coats than usual, and make their nests in the carcases.

The domestic mouse is considered to have originated in Asia, whence

(*right*) Wood
mouse washing

PLATE 21

(*below*) Common
dormouse

(*above*) House mouse

PLATE 22

(*left*) Grey squirrel

it has spread to every inhabited part of the world. Its early spread was doubtless largely unaided, but in more recent centuries, it may be presumed that it has travelled stowed away in stores and merchandise.

The head and body of the house mouse measure 3–4 in. (70–92 mm.), and the tapering, flexible and sparsely haired, scale-ringed tail is about the same. The weight ranges from ¼–1⅓ oz. (10–41 gm.). The snout is pointed, the bright eyes are black, the large, sensitive brownish ears are nearly half the length of the head, and the soft, brownish-grey fur is only a little paler on the underparts. Its voice is a well-known squeak. The house mouse is chiefly but not wholly nocturnal, active and silent in its movements, emerging from a tiny hole in floorboard or skirting and gliding without sound over the floor, ascending table-legs or walls with ease, and then, if alarmed, taking a relatively prodigious leap back to its hole. It is always a matter of surprise how small a hole a mouse can pass through. Wire-netting of not more than ⅜ in. (9 mm.) mesh must be used for mouse-proofing. Its climbing abilities, also, are considerable. Concrete floors will not keep it out of a house; it will climb the outer walls and enter the upper windows, thence making its way unnoticed to the lower floors behind woodwork or plastered walls, till it reaches the kitchen, the larder or the storeroom. Though it shows by its preference that its natural food is grain, it will eat practically anything edible. It can exist with little water, as was shown by mice that lived on flour alone in the 'buffer depots' set up by the government in World War II.

Success as a species is due as much to the mouse's fecundity as to its adaptability in the matter of food and shelter. Breeding is fairly continuous throughout the year. The number of litters a year, and the number of young in a litter, vary with habitat. Mice living in dwelling houses average just over five litters a year, with some five young to a litter. In the cold stores there are six litters a year with six young in a litter. In grain stores there are eight litters a year, and in corn ricks, where all conditions are highly favourable, there may be ten or more litters a year. Gestation takes nineteen to twenty days. The young are born blind and naked and are weaned at eighteen days. At the age of six weeks they begin to breed.

8

Records kept over the past fifteen years show that the rate of infestation in London is one building in every 100.

People sometimes find there is a difference between the ease with which house mice can be caught in traps in a country house as compared with the difficulty of trapping them in London. The explanation of this is, in all probability, that there really is a difference between town and country mice.

Originally, according to the experts, there were four wild subspecies of house mouse. Three of these have become commensal with man, and have given rise to a large number of other subspecies. The situation is further complicated by the way house mice are transported from one place to another by human agency, until the races are so mixed that even the experts tend to throw up their hands in despair.

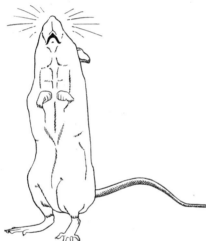

Broadly, there are three groups: those living in houses, those living in warehouses and on farms, and those that have gone back to the wild and live in fields. The first group finds conditions hardest because of continual disturbance by people and domestic animals. To have to struggle for a living may well bring out qualities of resource. On the other hand, there may have been genetic changes affecting behaviour.

The house mouse exhibits a considerable range of variation in colour, both darker and lighter than the typical 'mouse colour',

Aggressive attitude of a house mouse

and many of these variations may be the result of an attempt to breed 'fancy' mice. One example of this is the familiar white mouse, an albino with pure white fur, pink eyes, feet and tail. There are also dark, nearly black, and spotted variations. Several cases of mice that are hairless except for a few whiskers, have been recorded.

'Singing' mice are sometimes heard. The song may vary from a few

low soft notes to a song, sustained for ten minutes or
bird-like and resembles the more subdued vocalisation
Such mice always seem to be males and post-morten
them to be suffering from inflammation of the lungs or ⌐,
larynx.

Although the eye of a mouse is usually described as 'beady', suggesting a lustre associated with acute vision, tests showed that there was little response to movement of objects near at hand, whether in daylight or in half-light. The ears, on the other hand, will follow the course of such a moving object, suggesting that they are highly sensitive to vibrations in the air. It seems likely that the ears may be able to pick up ultrasonic sounds, and it has been suggested that mice may use ultrasonics as a means of communication with each other, or even as a form of echo-location for finding their way about in the dark.

A local subspecies of the house mouse formerly found in St Kilda had been dignified with the rank of species with the name *Mus muralis*. Its distinguishing features included less slender feet and tail, and slight peculiarities of the palate. Since the human inhabitants left the island in 1930, the St Kilda mouse has become extinct, and today it is regarded as having been no more than a subspecies, *Mus musculus muralis*.

Dental formula: i $\frac{1}{1}$, c $\frac{0}{0}$, p $\frac{0}{0}$, m $\frac{3}{3}$ = 16.

The Long-tailed Field Mouse

Family MURIDAE   *Apodemus sylvaticus*

The long-tailed field mouse, also known as the wood mouse, is as much an inhabitant of the field, the hedgerow and the garden as of the wood, but is rarely among heather or on high moors. It can bring something approaching despair to the owner of a kitchen garden, for its habit of taking newly sown peas (if these have not been rolled in red lead or soaked in paraffin), digging out bulbs, and so on. It also has a partiality for strawberries as soon as they ripen. The matter is made worse by its habit of storing foods of all kinds, which means that the amount of damage is out of proportion to its size and to its bodily needs.

The long-tailed field mouse is about $3\frac{1}{2}$ in. (81–94 mm.) from the long snout to the base of the tail, the tail being about the same length. Weight ranges from $\frac{1}{2}$–$\frac{3}{4}$ oz. (14–25 gm.). Females are slightly larger than males. The fur on the upper parts is a dark yellow-brown, the underparts are white. In adults the line of demarcation between upper and lower surfaces is always distinct. There is a spot of buff or orange on the chest, the varying development of which, in certain local races, has caused systematists to recognise four subspecies. The dark eyes are prominent, suggesting immediately that it is mainly nocturnal, and the long oval ears have the inner margin turned inwards at the base. The tail is dark brown above, and whitish below. It is the commonest of the British mammals in country places, but less frequent in Ireland. It is common in Europe as far north as Sweden and Norway.

As a rule the long-tailed field mouse uses runways in and under leaf litter. It constructs its burrow underground or under the roots of trees, and here it stores great quantities of acorns, nuts, haws, grain and smaller seeds for winter use, as much as a pint in one hoard, the acorns and other fruits being taken to the cache at the rate of one every half minute. The long-tail becomes less active in winter than at other seasons, although it does not hibernate. It comes indoors rarely, and then only in late autumn; its resemblance to the house mouse frequently leads to its being mistaken for that species. Usually it will not enter houses if house mice are present.

The breeding season begins in March, rises to a peak in July and August and ends in October or November. Breeding may continue throughout a mild winter. There may be several litters a year. Gestation is twenty-five to twenty-six days. The young per litter vary in number from two to nine, giving a high rate of increase, which is offset by the inroads in the numbers made by owls, foxes, weasels, stoats, hedgehogs and vipers. The young are weaned at twenty-one days, and except for those in late litters, will start to breed the same year.

A feature of the maternal behaviour, shared with many small mice, is that at times a female, when alarmed, will run with half-grown babies hanging to her teats, and keeping in step. She may leap a foot or so with the youngster still hanging on and taking no harm from the shock of landing.

The long-tailed field mouse is very quick in its movements, and when alarmed moves in a zig-zag manner. It will also progress in leaps, looking like a miniature kangaroo, but all four feet touch the ground together although the body is held semi-erect. It will readily climb bushes in order to obtain berries, leaping to the ground from considerable heights, or if seeking to escape capture, will leap several feet up into a twiggy growth and disappear. I once moved a pile of faggots under which was a company of these mice. When exposed they leapt in all directions, and one measured leap, from the ground on to a faggot, was 3 ft. (90 cm.) almost in the vertical. A long-tail has been recorded as jumping vertically from a height of 15 ft. (5 m.) and landing unharmed. These mice also swim well when the occasion arises. Accomplished burrowers, they often use unmortared stone walls for runs and stores, and are especially given to making their homes under stones or concrete blocks half-buried in the ground. They are gentle and apparently timid in the hand. Their call, a high squeak, is seldom heard.

These mice are gregarious, in the manner of those found under the pile of faggots, and the stores of food appear also to be communal. These may be, like the diet, very varied. Some of the items eaten or stored are leaves of clover and dandelion, with flower-buds of the latter, nuts of all kinds, apples, grapes, gooseberries, crocus and hyacinth bulbs, acorns, rose and bramble seeds, holly berries, slow-worms and eggs. Insects and insect larvae, as well as spiders, are also eaten. Long-tailed field mice have been known to enter beehives, and not only eat the honeycomb, but actually to nest there. On the other hand, there have been instances of the intruder being stung to death and the carcase completely enclosed in wax. A deserted bird's nest is often adapted to use as a feeding table, when seeking haws in the hedges, or as a permanent habitation, in which event the nest is roofed with moss.

Little is known of their social behaviour, except that the female drives the male from the nest as soon as she is pregnant. Also, when two long-tails meet they rear on their hind-legs and box for a few seconds, with very rapid movements of the fore-legs. A more ritualised behaviour has been recorded, when three 'eggs' were seen in the grass, arranged in a Y. Two of these reared up, to reveal themselves as long-tails. Standing on their hind-feet, they put their front paws on each other's shoulders. With noses almost touching they sniffed, then each nuzzled the other, after which both dropped to earth again, to become two brown 'eggs' once more. After a few seconds' pause one of them again reared, as did the third, to repeat the performance. So it went on, in twos, turn and turn about, after which the three field mice went their separate ways.

'Sniffing noses' is a common social trick in rodents, and I have seen a field mouse and a dog, coming face to face on a path, sniff noses, then each turn and go its separate way.

The breeding nest is a ball of dry grass, usually built in a separate chamber of the burrow, but it may be above ground in a heap of vegetable rubbish. Some burrows may extend as much as 3 ft. (90 cm.) underground.

The stir caused in south Devon a century ago has recently been recalled. There was an overnight fall of snow, and small hoofprints were seen going over the ground, under bushes, and also over roof-tops and haystacks. The tracks ran for an estimated 100 miles. In one place they went up to a wall and started again on the other side, as if something had gone clean through. Many explanations, prosaic or fantastic, have been offered in the interim to explain the 'Devil's Hoofmarks' and suspicion now points to long-tailed field mice. In 1964 Mr Alfred Leutscher suggested that the supposed mysterious hoofmarks might have been the tracks of long-tailed field mice.

These mice usually run over the ground but may progress by leaps, landing each time with all four feet together to leave a U-shaped impression, $1\frac{1}{2}$ by 1 in., at 8 in. intervals, precisely the dimensions recorded for the trails seen in south Devon in February, 1855.

If his suggestion is correct, why has it taken over 100 years for somebody to spot the real culprit? Field mice are common enough, but few people have paid much attention to their tracks. We have had to wait for the time when three things coincided: somebody who made a special study of them, easy methods of photography, and, most important of all, the moment when the penny drops and that person, seeing the U in the snow, recalls the story of the 'Devil's Hoofmarks'.

Yet mystery still remains. Why, if field mice are so common, are trails of miniature hoofmarks not more familiar? Why only in south

Devon? The truth is that other trails of this kind have been seen and given rise to speculation during the century that has elapsed since the south Devon affair, but because they were not so numerous and did not hit the headlines they have been overlooked. The south Devon trails may very well have been the result of an unusual population of field mice similar to the one that occurred in Yugoslavia and which is dealt with under the next species, the yellow-necked mouse. This, together with a light fall of snow, could furnish the rest of the explanation.

Occasionally a small pile of stones may be found covering, in the manner of a cairn, the entrance to a vertical tunnel into the ground. It is the work of the long-tailed field mouse. I have heard it said these 'cairns' are common, but in my experience they are rare.

The first one I saw was made of small pebbles and a few pieces of twig. The latter seemed to be supporting the stones as if placed deliberately to prevent them falling into the hole. The stones were so arranged that they formed a sort of igloo with an entrance through which the mouse could reach the mouth of the tunnel. Live-trappings showed that at least six mice were using the tunnel, but there was nothing to suggest that they had combined to build the cairn. In due course the stones were removed and scattered over the bare earth. Because the cairn had been photographed it was possible to recognise the individual pebbles and twigs and to know for certain that these were those that had been scattered. Presumably it was done by one or other of the mice.

The next cairn in my experience was dismantled because it was on a lawn and the grass had to be cut. The stones and twigs were kicked aside and the following morning they were all back in position just as if they had not been disturbed. The third cairn I was able to study did not form a roof over the entrance of the hole, but was arranged in a neat pile immediately to one side, one edge of the pile being along the rim of the tunnel entrance.

The purpose of the cairns is unknown and since it seems that only a small minority of the millions of field mice tunnels that one comes across are treated in this way it is difficult to form a hypothesis to account for them. Some of the pebbles are relatively large: the largest I weighed measured $\frac{1}{3}$ oz. (12 gm.), half the weight of the mouse that transported it. Although the following incident does not assist the explanation, it suggests a probable line of enquiry. A field mouse was kept in a large vivarium with several inches of soil on the bottom into which it burrowed. In excavating the earth was thrown aside in the form of pellets about $\frac{1}{4}$–$\frac{1}{3}$ in. (7–10 mm.) in diameter and these were strewn around the cage. On one occasion the mouse was seen leaping backwards and forwards between a point just in front of the mouth of its burrow and another point 6 in. (150 mm.) away. This it did repeatedly for two to three minutes. Each time it leapt back to the mouth of the burrow it carried an earth pellet in its mouth which it deposited and pushed down with its fore-paws. Altogether a dozen pellets were moved, the mouse resting for a short while between each jump; had they been pebbles, there would have been constructed one of these puzzling cairns.

Forms related to the long-tailed field mouse are the Hebridean field mouse (*A. sylvaticus hebridensis*), with the white of the underparts dotted with buff; the Fair Isle or Shetland field mouse (*A. sylvaticus fridariensis*), like the yellow-necked but without the collar; St Kilda field mouse (*A. sylvaticus hirtensis*), with brown underparts; and Bute field mouse (*A. sylvaticus butei*), darker, with shorter tail and ears. The status of these subspecies is, however, in some doubt and the authorities are somewhat divided upon which to recognise as valid subspecies.

Dental formula: i $\frac{1}{1}$, c $\frac{0}{0}$, p $\frac{0}{0}$, m $\frac{3}{3}$ = 16.

The Yellow-necked Mouse

Family MURIDAE   *Apodemus flavicollis*

Towards the end of the last century, Mr de Winton called attention to what was considered to be a new British mouse, the yellow-necked mouse, distinguished from the long-tailed field mouse by its larger size, the head and body measuring 4–5 in. (85–124 mm.), and by the brown spot on the chest (commonly found in the long-tailed field mouse) being extended into an orange cross whose arms connected with the upper side colouration, which is described as golden brown. There used to be a division of expert opinion on whether or not this is a species distinct from the long-tailed field mouse. On the Continent the two appear to differ in habits and habitat. In this country the differences are less marked. Nevertheless, it is now accepted that the two represent different species. The yellow-neck is most common in southern England, becoming rarer to the north. It is not found in Scotland, the Isle of Man or Ireland.

Since there are variations in the amount of yellow or orange on the chest it is not always easy to tell a long-tail from a yellow-neck and the juveniles of the two species are very much alike. Two marked features will aid identification: yellow-necks are much more given to entering houses in autumn, and adults are more robust in appearance, more energetic and better jumpers than long-tails.

Whoever wrote 'Quiet as a mouse' could not have been thinking of yellow-necks. Each autumn they invade my house. They scramble up through the cavity walls and the noise they make is quite considerable. They race behind the skirting and under the floorboards, and when one of them is heard, in the quiet of night, bounding between the joists the sound is like that of heavy human footsteps. It is highly probable that yellow-necks may be responsible for some reports of ghostly footsteps.

My yellow-necks raid the apple store in the cellar. They also bring in nuts from hazel, filbert and walnut trees, although most of these trees are about a hundred yards from the house. The shells can be identified, so we know they transport the nuts over these distances, but so far nobody has caught one in the act. Moreover, they take the nuts up the cavity walls to the third floor where, under the floorboards, there are vast collections of empty shells, the accumulation of years of tremendous labour on the part of the yellow-necks.

This species was involved in another astonishing incident, but on a larger scale. In August, 1967, the Press reported a nine-mile-long stream of mice advancing across Bosnia, in Yugoslavia. The mice were said to be so arrogant and aggressive that the cats dared not attack. Damage to crops was estimated at £190,000. A later report described the mice as six inches long and yellow in colour. This made their identity somewhat puzzling and the suggestion of large columns advancing over the countryside did not fit in with previous knowledge about the behaviour of plagues of house mice, which have often been reported from such places as California and Australia. Nor were they lemmings for they do not occur in southern Europe.

It later transpired that the species involved were long-tailed field mice, and yellow-necked mice. The latter, which may measure as much as 5 in. long, often look yellow when, as sometimes happens, the bright yellowish-buff collar spreads along the lower flanks.

The cause of this plague was that the previous winter had been mild, with abundant beech mast and acorns. This increased food supply and the mild weather had eliminated the usual high winter mortality, had led to a higher rate of breeding in the following spring and so had caused a population explosion. By the summer both kinds of mice were spreading out from the woodlands on to the arable land. Such were their swarms that they even entered farm-houses and buildings, which long-tails seldom do and yellow-necks usually do only in winter.

Reports of plagues of house mice are not uncommon. These follow bumper harvests in other parts of the world, notably Australia, when granaries are full. Plagues of meadow voles are also the result of an excess of food, in this case grass. In neither of these do we have suggestions of directional migrations, that is, of house mice or meadow voles advancing in columns or phalanxes. This is a familiar feature of the stories about lemmings, but even here investigations in the last few years have revealed little more than a spreading out from centres of abnormally increased populations.

The reference to a nine-mile horde in Yugoslavia suggested an army of mice on the move, but this, as investigators on the spot later found, was due partly to the imprecise use of words, although it may have some foundation in fact. For example, several years ago when a field of wheat adjacent to my garden was cut, rats invaded my kitchen garden in large numbers. They sampled every kind of vegetable, even climbing into the tops of the brussels sprouts and eating them down to the base. It is significant that a neighbour reported having seen 'a mass of rats' making for my boundary fence. The rats suddenly departed after two days and then we heard reports of people in the village having seen 'an army of rats moving along the banks of the stream'. No doubt in all plagues of this kind there are limited and local directional movements such as these. In Yugoslavia the mice invaded the farm-houses in large numbers. They may have come in by one door and out by another, so giving an additional impression of an army on the move.

The report that cats were afraid of the mice appeared somewhat laughable to many people when the first reports appeared in the Press, but yellow-necks and long-tails are robust and active mice that leap about in all directions when disturbed, like swarms of giant fleas. A cat faced with this could very well settle for another kind of food and move away, not so much frightened as bewildered.

Possibly, also, the mice find strength in numbers. This has been noted in other fields of natural history. It was evident in the rats feeding in my kitchen garden. I could walk among the rats and watch them feeding, hear them filling the air with a nauseating sound of chewing, and they took no notice whatsoever of my presence. Either they had grown up undisturbed in the cornfield, during the months the corn was growing and so failed to recognise man as an enemy, or else their serried ranks gave them confidence.

The density of the Yugoslav mice was on average four per square yard. This, continued over a large area, would give an exaggerated impression, as of an army. In addition, small animals always look more numerous than they really are when moving about rapidly.

Dental formula: $i \frac{1}{1}$, $c \frac{0}{0}$, $p \frac{0}{0}$, $m \frac{3}{3} = 16$.

The Ship Rat

Family MURIDAE   *Rattus rattus*

The terms 'black rat' and 'brown rat', commonly used to distinguish the two species, tend to cause confusion because the 'black rat' (*Rattus rattus*) is often brown and because melanic variants of the 'brown rat' (*Rattus norvegicus*) also occur. It is becoming the practice to speak of *Rattus rattus* as the ship rat (alternatively roof rat or house rat) and *Rattus norvegicus* as the common rat (alternatively Norway rat or sewer rat). Surveys made of the distribution of the ship rat, in 1951, 1956 and 1961, show it to be confined almost exclusively to ports and to a few towns (such as Ware and Slough) which are connected to sea-ports by canal. The 1956 survey showed that the ship rat was 'always present somewhere' in forty-eight localities in the British Isles, but by 1961 this total had been reduced to twenty-eight. This decline is largely due to new buildings, more efficient rat control and legislation to enforce de-ratting of ships.

Both ship and common rats are of Asiatic origin, although the ship rat is thought to have been settled here for six centuries before the common rat.

The ship rat is of more slender proportions than the better-known common rat. Its head and body together measure 6½–9 in. (165–198 mm.) long, while the scaly-ringed and almost hairless tail is up to 10 in. (254 mm.) long. The weight is variable, up to about ½ lb. (200 gm.). The snout projects far beyond the short lower jaw; the whiskers are long and black. The naked ears, contrasting with the finely haired ears of the common rat, are the best means of identifying the live animal. The feet are pink, with scale-like rings on the undersides of the digits and five pads on the sole. The thumb of the fore-feet is reduced to a mere tubercle. The ship rat is more of a climber than a burrower, and cleaner in its feeding than the common rat.

Although rats are associated with garbage and are generally offensive to us in their habits and appearance, they maintain their own cleanliness, spending much of their time cleaning their fur and paws. In Mediterranean countries and India, where the ship rat lives a more outdoor life, it climbs trees and mostly makes its nest in them. In this country the doe collects quantities of suitable materials, such as rags, paper and straw, and constructs a roomy nest, using this for the three to five litters a year. Breeding occurs throughout the year with a peak in summer and, less often, in autumn. Seven is usual, but litters may vary from five to ten. The gestation period is twenty-one days. The pink-skinned young are born without fur, sight or hearing. They become sexually mature at three to four months.

Both ship and common rats are omnivorous; anything that can be digested is eaten. Cannibalism no doubt occurs under unusual conditions, but the idea that the common rat is the more ferocious and has driven out the ship rat by aggression is almost certainly erroneous. The replacement of the ship rat, which prior to the eighteenth century was spread across the British Isles, is more the result of competition for

food and living space. In rural areas in some parts of the tropics the two
species share the same habitat, and generally in the tropics the ship rat
is the more successful and has not been ousted by the common rat
except, surprisingly, in some ports.

The ship rat has been divided into three subspecies: *R. rattus rattus*,
pure black above, black or dark grey beneath, *R. r. alexandrinus*, brown
above, grey beneath, and *R. r. frugivorus*, brown above and white or
cream beneath. It has been suggested, however, that all are polymorphic
forms of one species and in Britain all three forms seem to live together
and to inter-breed, and all gradations exist between the white- or cream-
bellied *frugivorus* and the typical black *rattus*. In London about 70%
of ship rats are of the *frugivorus* type and only 3% are positively
alexandrinus.

Dental formula: i $\frac{1}{1}$, c $\frac{0}{0}$, p $\frac{0}{0}$, m $\frac{3}{3}$ = 16.

The Common Rat

Family MURIDAE   *Rattus norvegicus*

The natural method of dispersal for mammals is that the more
adventurous individuals make food-finding excursions beyond the
district in which they were born. In this type of spread climate, mountain ranges, broad rivers or seas usually act as checks on further progress.
Rats have kept close to man. Where man is, there is food and shelter
whether for himself or for his domesticated animals. When man loads
his ships with grain and other food stuffs rats have tended to go with
them. They even get into bales of merchandise and are conveyed in the
holds of ships; or, failing that, there are always mooring ropes to serve
as bridges from the quay to the vessel. The result has been that, over
and above the original natural spread, rats have been carried unwittingly
by man to all corners of the globe. The arrival of the common rat in
this country was only one small stage in the process.

The origins of the common rat in this country are uncertain. It is
thought that it spread across Europe from central Asia and reached
England in ships coming from Norway at about the time the Hanoverian
kings ascended the throne of England. The alternative names of the

common rat are, as a consequence, the Norway or Hanoverian rat. The general concensus now seems to be that it did not come from Norway and that the reference to Hanoverian kings was more of a political joke. The alternative view is that it was brought here by trading vessels from Persia, or farther east, between 1728 and 1730. There are reports of rats crossing the Volga in large numbers following an earthquake, in 1727, but it is unlikely these would have reached this country in three years or less. It is fairly well substantiated that the common rat reached Paris in 1753, which is further evidence that it did not reach England by the overland route.

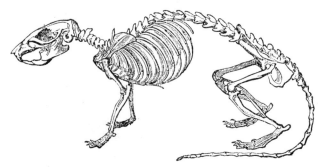

Skeleton of common rat

The common rat is more heavily built than the ship rat. The combined length of its head and body is 8–10½ in. (203–267 mm.), while the thicker, scaly-ringed tail is only equal to, or less than, the length of the body alone. Weight is variable, up to 1½ lb. (500 gm.). The head is proportionately shorter, the ears smaller and covered with fine hairs, and the eyes smaller and more prominent. The fur on the upper parts is typically grey-brown with a tawny tinge, and is more shaggy than that of the ship rat. On the underparts it is a dirty white. The feet are flesh coloured and the tail is often bi-coloured, paler underneath than above.

The common rat is sexually mature at the age of three months, and thereafter produces three to five litters in a year. Ordinarily these comprise four to ten blind, deaf and naked young; but much larger litters are on record. They leave the nest at about three weeks of age.

In the country, where the species is known as the barn rat, its chief enemies are the tawny owl, stoat, weasel, fox and, of course, man. The stoat, weasel and fox do not always receive appropriate credit for their services. Admittedly the three also destroy poultry, but there is no question that the damage done by rats is much more serious. The mischief they do in chicken-runs is less spectacular but more extensive than anything their predators can do. Being good swimmers and divers, even ducklings afloat are not safe from them. Rats are equally destructive to property, livestock and stored crops, especially such root crops as sugar beet in East Anglia, in clamps. Scores of them will burrow through the cover of earth, damaging the stored roots and fouling more than are eaten. It is in such situations that weasels are most effective as destroyers of rats. Hawks and foxes come into play more in the open countryside, where the rat rarely ventures far from cover of some sort.

The versatility and the destructive power of the common rat lead it to spoil and waste man's food stores and do extensive damage to property. At one time it performed its natural service of scavenger in this country, and still does elsewhere in the world, but as soon as man has learned to attend to this for himself, the rat becomes a mere parasitical nuisance, sufficiently omnivorous for there to be no possibility of starving it out; so a costly war must be waged on a pest already expensive to us. Unfortunately, rat-killing campaigns that do not cover every square yard of the country can only have the effect of temporarily mitigating the pest. Even when determined onslaughts are made, as used to be in the so-called 'rat weeks', the rat's fertility is so great and its recovery so rapid that the loss of $\frac{9}{10}$ of a generation is quickly made good.

It is often stated that the common rat is thoroughly omnivorous. Although it will eat greenstuff, it is basically a grain eater, infesting grain stores or wheat and oat fields when the grain is ripening. Although highly successful because it has become a commensal of man, the common rat is cautious of and avoids new objects in its environment. This makes it trap-shy. It is also bait-shy. When it has had a sub-lethal dose of poison it may remain bait-shy for up to six months.

One of the more obvious differences between the common and the ship rat is that the latter is more of a climber. This must be qualified. In my own garden there was a time when common rats had a regular run under a yew hedge on my boundary into a cornfield. When a neighbour's cat developed the habit of coming regularly to sit by the run the rats ran along the top of the hedge. They even turned the top of the hedge into a playground, making platforms of deciduous leaves on it. They took to climbing into adjacent holly trees and, most spectacular of all, they would run along the top of a split-chestnut fence, to reach the yew hedge, bounding from the top of one spiked stake to another, with a dexterity and speed that had to be seen to be believed.

From time to time one hears tell of somebody who has encountered a column of rats migrating more or less in formation. Little is known about these mass movements of rodents, largely because they are always seen by untrained observers, whose accounts are not strong on the details. This subject is dealt with more fully under the yellow-necked mouse (q.v.).

With accounts of rats migrating in columns, stories are sometimes linked of one of the rats carrying a straw in its mouth, with another rat holding on; or it may be that the story is of just two rats, one hanging on to the straw. In all such instances it is questioned whether this is a case of a blind rat being led by another.

One does occasionally come across other reports suggesting that some of the higher animals are capable of good samaritan acts, but it is difficult to credit a rat with any form of compassion. There is, however, a more likely explanation here. It does sometimes occur that a rat will seize a piece of food and run off with it. Another rat seizes the food and the two are seen running in tandem. On rubbish dumps rats will carry off small bones, and again it can happen that a second rat will seize the other end of the bone.

Rats in cornfields will sometimes run off trailing a straw in the mouth. It needs only a second rat to seize such a straw to give rise to the impression of a blind rat being assisted by one of its fellows. It may be that this could happen at the time when a colony of rats was deserting a former feeding ground, so giving the appearance of a compassion act.

Rodwell, writing in 1858, states that the first person to breed white rats (the albino of the common rat, not of the ship rat) on a commercial basis was a man named Ostin, who obtained a couple from Normandy. From these he bred a large number which he sold for four shillings a pair in London.

PLATE 23 (*above*) Fox's earth (*below*) Badger's set

(*above*) Year-old dog fox

PLATE 24

(*left*) Three-month-old fox cub

Twenty years ago I became interested in the story, about which many people are sceptical, of two rats combining to carry away an egg—and at last I have a possible explanation.

The story is of one rat being seen on its back clutching an egg while another drags it along. On this basic theme there are several variations: that the recumbent rat is dragged by the tail, by a leg, by an ear or by the other rat gripping some part of the head with its teeth.

On the face of it the story is improbable, particularly as there is no other evidence of rats co-operating in this or any other way, yet a considerable number of sincere, intelligent people from varying walks of life claim to have seen it happen. That is what gives the story an air of verisimilitude.

I have long assumed the story to be essentially correct and have tried all known means to see it for myself. To that end I have many times kept watch where I knew rats to be, putting down eggs to tempt them to perform, but although care was taken to remove human scent from the eggs in all instances the rats ignored the bait—usually the egg disappeared soon after I gave up keeping watch! When I have found rats removing eggs, potatoes or the like I have, when practicable, dusted the surrounding ground with flour or chalk, so that if a rat were dragged along on its back, for any reason, there would be a tell-tale mark, as if one had made a path through the powder with a medium-sized paint brush. So far the only marks I have seen have been numerous rats' footprints or tail marks, all quite normal.

I enquired of everyone I met, who dealt with rats, whether they had seen such an incident. Several years ago I heard of a rat-catcher who had recounted this story as a personal experience. I arranged to call the next time he was due, and casually got into conversation with him and referred to this well-known story. He told me how he had opened the door of a shed one day and saw a rat scuttle away from near the threshold. He also saw, at the point from which the rat had run, a piece of sacking with an egg on it. Then, to his surprise, the sacking turned over, the egg rolled away, and he realised the 'sacking' was a second rat. He assumed he had disturbed two rats in the act of co-operating to transport an egg.

On a number of occasions I have been called to inspect hens' eggs discovered a spade's depth beneath the ground, usually when potatoes are being dug. Tunnels through the nearby earth, directed towards the neighbouring hen-run, seemed to indicate that rats had carried the eggs there. There is also evidence of rats stealing eggs in other ways. Since the eggs are undamaged and show no teeth-marks, there is no indication of how they were transported.

I have tried to get a line on this for squirrels, rats and mice, so far with little success. Recently, another 'rat man' came to see me. We talked about a story he had heard of two rats and an egg but about which he was highly sceptical. We also discussed the way rats carry eggs and he assured me he had once seen a rat carrying an egg held under the chin and supported by the front paws, the rat bounding along on its hind-feet. This seemed improbable, but he assured me he had seen one egg and several apples and potatoes carried that way.

It was then that an idea occurred to me. One of the well-known tricks of behaviour when two rats meet is for one to fall on its back, in what is called a submissive attitude. The other rat proceeds to groom the recumbent rat, licking the fur, especially in the head region.

It is not easy to obtain precise details, but those specialising in the behaviour of wild rats are emphatic that the submissive attitude, under the circumstances described here, is a fairly frequent occurrence and that the grooming may extend to other parts of the body.

Suppose that we can accept the possibility that a rat may occassionally carry an egg, using its fore-paws. Suppose also that a rat engaged in this meets another and falls on its back (still retaining a hold on the egg) and the other proceeds to groom it, the picture presented would come very near to that so often described: of one rat on its back holding an egg and the other *apparently* pulling the first rat. The appearance would be sufficient to deceive an untrained observer, and the verbal description, slightly exaggerated as eye-witness descriptions tend to be, would account for the rest of the story.

This is highly speculative, and there are several gaps in the argument. For example, we yet need incontrovertible proof of eggs or other objects being carried with the aid of the fore-paws. We also need proof that the grooming extends to parts of the body other than the head region, but 50% or more of orthodox scientific knowledge is speculative. Speculation is the precursor of hypothesis, and it is by testing hypotheses that firmer knowledge is often gained.

The submission, egg-carrying and grooming could reasonably have resulted in the description of one rat scuttling away from an egg on a piece of 'sacking'. The episode was over in a second, and anyone with more imaginative inclinations, seeing a similar event and adding a few embellishments, could easily retail the more familiar version, of one rat dragging the other along.

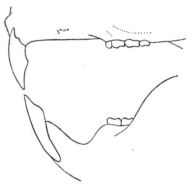

Teeth of common rat

Statistically there is no objection to this idea. There may be a very slim chance of seeing what must be an occasional and transitory event, but there are tens of millions of rats in Britain alone, and tens of millions of eggs and a similar number of human beings. The chance of all the circumstances coinciding, for example once a year, would be sufficient to account for the firmly established nature of the story.

Dental formula: i $\frac{1}{1}$, c $\frac{0}{0}$, p $\frac{0}{0}$, m $\frac{3}{3}$ = 16.

The Red Squirrel

Family SCIURIDAE   *Sciurus vulgaris leucourus*

The red squirrel is one of the most attractive of our small mammals, especially when it is sitting on its haunches on a branch, its feathered tail curled up its back, its tufted ears erect, holding in its fore-paws a nut, or when leaping with characteristic grace from bough to bough.

From the end of the snout to the tip of the tail proper, that is, excluding the hairs that extend beyond the tip, it measures only about $15\frac{1}{2}$ in. (390 mm.), and of this nearly half is tail. Weight ranges from 9–12 oz. (260–345 gm.). It is customary to speak of the tail as bushy. Rather it should be described as feathered, the hairs extending outwards and backwards on each side. The muzzle is well equipped with whiskers, the prominent eyes are black and bright, and the large pointed ears bear tufts of long hairs in winter. The hind-legs are much longer than the fore-limbs, and the heel of the long hind-foot is planted on the surface of the bough when the squirrel is at rest. The feet are well adapted for climbing. The fore-feet have four fingers and a rudimentary thumb, and the hind-feet have five toes. The soles are hairy, and the long curved claws are needle-sharp. The upper parts and tail are brownish-red and the underparts white. Before the winter, when the fur becomes softer and thicker, a grey tinge is developed on the sides, and the ear-tufts

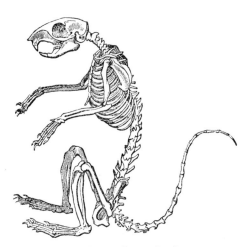

Skeleton of a squirrel

become longer and bushy; these are shed in the breeding season (early summer). The tail may have a creamy tint in summer.

The squirrel is diurnal, with peaks of activity in the early morning and just before dusk. In former times it was wantonly persecuted squirrel hunts being a popular sport in areas where it is now no longer found. Groups of men and youths would stone it from tree to tree until they had forced it into one that stood alone. There it was stoned until, in an effort to escape it dropped to the ground, usually to succumb under a shower of stones. Some years ago the squirrels that added to the attractions of Richmond Park were shot by the keepers to prevent them being killed in this way by gangs of youths coming from London. As one writer has put it 'it would have been better to discipline the offenders.'

The squirrel builds nests in the branches of trees it frequents, not merely for a nursery, but for other purposes. There may be several of these in adjacent trees, and some may be crows' or magpies' nests converted by the new tenants, but usually they are the squirrel's own work. They are bulky structures composed of twigs, strips of thin bark, moss and leaves; some are cup-shaped, others domed. These are usually known as 'dreys', but in parts of Surrey they are 'squaggy jugs'. The breeding nest is a huge ball of sticks and leaves, sometimes made in a

Red squirrel asleep, slumped across a branch

roomy hollow in a tree trunk. The breeding period largely depends on the quantity of food available and on weather conditions. Typically there are two main seasons in the south, January to April and late May to August, but apparently only one in Scotland. The gestation period is about forty-six days. Litters may contain one to six young, the usual being three or four, born blind and naked. They remain with their parents until they themselves are adult. They are weaned at seven to ten weeks. They develop hair at one week, the lower incisors erupt the third week, and the first set of teeth is complete by ten weeks. During the fourth and fifth weeks the eyes and ears open. There is a first moult at about fifteen weeks, a second moult at seven months and sexual maturity is probably reached between six and eleven months.

The drey is a spherical mass of sticks situated on the branches of a conifer close to the trunk. It is lined with leaves, moss, grass or bark, and it is smaller and more compact than that of the grey squirrel. There is no separate summer nest, but sometimes the young may be born in a hollow in a tree.

The food of the squirrel is fairly varied. In pine woods the cones provide the staple food, and the ground beneath a squirrel's tree will be found littered with the scales and cores of the cones from which the seeds have been extracted. This litter is the most tell-tale sign of the presence of a squirrel in the vicinity, and this is true for both red and grey squirrels. In beech woods they feed largely on mast, the sharp-edged triangular seeds contained in the prickly beechnuts. The hazel-copse also receives

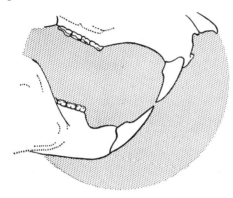

Teeth of red squirrel

attention, the nuts being a favourite food for storing in considerable quantities in holes in the ground. Wild cherries are also taken. An item in a squirrel's diet which perhaps results in the most damage is the bark taken from the leader shoots of young pines, thereby deforming the trees. This is offset, to some extent, by its feeding on the larvae of the pine tortrix moth which burrow in the tips of the young pine shoots. It also takes birds' eggs and nestlings.

The red squirrel does not hibernate. In the winter it may sleep more than in the summer, but it is still active, even during cold weather. Although typically arboreal, its habitat being woodlands rather than parklands, it feeds a good deal on the ground at times, especially on wild strawberries, bilberries and fungi. It is, moreover, an expert swimmer, although it only takes to water deliberately in exceptional circumstances. Popular belief credits it with spreading its tail to act as a sail, which probably means that it merely keeps the tail out of the water.

The voice is a rasping chatter followed by a hoarse call or whine. The young have a shrill piping call.

The squirrel was formerly distributed throughout Great Britain and

Ireland, wherever there was sufficient woodland, but it is now absent from the Isle of Man and absent or, at best, scarce in the south-eastern counties and in a large part of the Midlands; its main stronghold in England now is East Anglia. A survey made in Scotland in 1957, covering 202 forests, showed no red squirrels in 77, a scarcity in 109 and an abundance in only three of the areas. Outside Britain, it extends across Europe, from the tree line southwards to the Mediterranean, and into Asia. Throughout this area a number of subspecies are recognised. The British red squirrel is one, its subspecific name (*leucourus*) referring to the whiteness sometimes seen in the tail.

Dental formula: i $\frac{1}{1}$, c $\frac{0}{0}$, p $\frac{2}{1}$, m $\frac{3}{3}$ = 22.

The Grey Squirrel

Family SCIURIDAE   *Sciurus carolinensis*

The grey squirrel is a native of eastern North America, introduced as a pet in the nineteenth century and persecuted as a pest in the twentieth century. It had been liberated in various parts of the country at different times, but made no spectacular progress until, towards the end of the last century, it was introduced at Woburn Park, Bedfordshire. By the beginning of World War I it was spreading from that county into Buckinghamshire and Hertfordshire. Some of the squirrels from Woburn were liberated in the London parks, as well as in other parts of the country. Once the spread had started it continued, mainly to the west and to the north. Now the grey squirrel is established along the south coast, from Kent to Dorset and Cornwall, and northwards to the Scottish border. It appears to be absent from the Isle of Wight, Anglesey and the Isle of Man and also Cumberland, and there are only sporadic records for Norfolk, Caernarvonshire, Westmorland and Northumberland. There are areas of infestation in eight counties of Scotland, and some introductions into Ireland.

Larger than the red squirrel, it is a speckled grey above and white on the underparts, the head and body measuring nearly 1 ft. (30 cm.), the tail 8–9 in. (20–24 cm.). Its weight is 18–20 oz. (510–570 gm.). Apart from slight tufts in summer, it lacks the pronounced ear-tufts of the red squirrel and the tail is markedly less bushy. Its habits are similar to those of the red squirrel except that it is more aggressive, and it frequents open woodlands and parklands rather than the coniferous woods, one result of which is that it is not found in such high altitudes as the red squirrel.

Reports of red squirrels seen in areas where only grey squirrels are known to be are not infrequent. Changes in the coat in both species, throughout the year, cause some confusion. Red squirrels have more grey in winter and grey squirrels more chestnut in summer. There is also the complication of the grey having ear-tufts in winter.

Nevertheless, there should be no mistaking the red, particularly in its winter dress because of the very pronounced ear-tufts.

The false reports arise as a rule from occasional grey squirrels in which the darker patches on the body are replaced by reddish hairs. These erythristic forms are also responsible for some reports of interbreeding between the two species. There is no instance of hybrids between red and grey in the wild and, I believe, none in captivity.

In addition to the foods one normally associates with squirrels, the acorns, nuts and beech mast, grey squirrels eat a variety of things. They eat toadstools, as well as the *Boletus*, and they will eat puff-balls, not infrequently burying these fungi. They will tear oak galls to pieces, presumably to eat the insect larvae. Several times a grey squirrel has been seen carrying the carcase of one of its fellows, which has led to the argument whether this was some sort of compassionate act, a kind of rescue attempt, or whether the squirrel was transporting the carrion to eat it.

It is always hard to interpret occasional events of this kind. To turn to another we may wonder why, suddenly, in a public garden the squirrels, with adequate food around and plenty of wooden objects upon which to exercise their teeth, should chew to pieces the lead labels naming the trees. One finds, also, that for a while one individual squirrel will take to biting off whole leaves of horse-chestnuts, gnawing near the base of the petiole so that the ground below becomes littered with whole green leaves.

The leaf-cutting may be connected with sampling the sap, as is some of the barking, for example, taking the inner layers of the bark of sycamore and beech for the sugars. Another trick is to strip bark from the dead twigs of limes, but this is used for lining nests. In these and other ways, as when the squirrels eat the green buds of young trees, squirrels are unpopular with foresters. Beech especially is liable to attack and heavy damage can result, but the damage is sporadic. Some beechwoods may show little sign of it, and of those affected not all the trees are damaged. It is now suspected that this is linked with territorial behaviour of the females and the lines of damaged beech trees often mark the boundary between the territories of two dominant females.

The males show no such behaviour. Although they help build the nests they are driven out when the female becomes pregnant, and she defends the nest and the immediate vicinity against intruding squirrels, often with considerable violence.

On the rare occasions when squirrels make hoards they use cavities in trees. Normally nuts, acorns and other foods are buried singly and well spread out. Buried food is found by scent, even when the ground is covered by snow, so that an individual may not necessarily return to the nut or acorn it has buried but all squirrels in the locality share the cached foods. Often the food may be carried some distance. In 1963, when there was a heavy crop of horse-chestnuts and a shortage of other nuts in Surrey, tremendous quantities of conkers were transported and buried, some of them being carried up to two hundred yards, the trails being marked by a thick carpet of husks. All the conkers, numbering thousands, were removed from under one clump of chestnut trees. Some were buried in the immediate vicinity, but most were carried along one route, which was marked by the husks, and then buried spread out over the field.

During the pre-mating period it sometimes happens that several males will chase one female until the dominant male drives the others away. There are other variations on this theme, but the most spectacular I witnessed was in January 1950, on Sheen Common, then on the outskirts of London. My attention was drawn to a barking note being made by several grey squirrels coming through the bracken. They converged on a large oak into which nine or ten of them climbed. It was difficult to be sure of the precise number because once in the oak they ran round its branches in a sort of follow-my-leader, creating what looked at times like a large, irregular revolving circle. Round and round they went over the branches and up and down the trunk for about a minute, after which they left the tree singly or in pairs, dispersing in all directions as quickly as they had assembled.

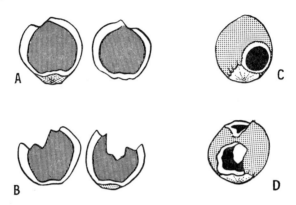

Nuts that have been opened by various rodents:

A   Neatly split in half by an adult squirrel

B   Split after gnawing by young squirrel which has had little practice at nut-cracking

C   Neat, round hole made by dormouse

D   Untidy holes made by long-tailed field mouse. Note domage to surface of the shell surrounding the holes, which are made by the upper incisors

The sound made by males chasing females is said to be low and vibrating like the song of a grasshopper. Barking is the response to danger, but then it is very rapid and as danger recedes it slows down to the familiar scolding which has been rendered *chuk-chuk-chuk-quaa*. The bark that preceded the oak tree incident was not like either of these but consisted of isolated single barks. However, grey squirrels use many variations of purring (when feeding), chattering, scolding and barking.

Two kinds of nest are made. One is the winter drey, also used for a nursery. It is of leafy twigs, domed and usually in the angle between a branch and the trunk. It is lined with leaves, bark, moss or grass and especially honeysuckle bark. The summer drey is a leafy platform built out on the branches.

The breeding season lasts from the end of December until June. Gestation is forty-four days. There are two litters a year, of one to seven young, but the usual number is three, naked and blind at birth. Weaned at seven to ten weeks, the young reach adult size at eight months and are sexually mature at six to eleven months. The life-span is possibly five to six years.

Squirrels have long been renowned for their playfulness, but whether their behaviour is play in the true sense of the word is a matter of opinion. Both in the wild and in captivity they frequently indulge in a wide variety of rapid acrobatic evolutions, sometimes using a fir cone or stick as part of the performance, which in my experience is most often to be seen about an hour after dawn in the summer.

A typical instance was a wild squirrel which I watched passing a twiggy stick rapidly backwards and forwards between its incisors, as if tasting it. Then it took the stick from between its teeth and, using the fore-paws almost as hands, swung the stick round and put it back between its teeth. It executed a backward somersault and followed this with other acrobatics as it twisted and turned on to its back, twirling the stick with as much skill as a drum-major. It leapt into the air, turning to fall on its back, leapt up again to land on all fours, somersaulted, and finally raised its hind-quarters to kick out with its hind-feet. Twisting, turning, somersaulting, bucking on all fours the squirrel gave a magnificent high-speed display. In the end it dropped the stick, made a standing leap into the air, bounced across the grass on stiff legs, leapt at the base of a tree and streaked up the trunk.

Endless variations on this theme may be carried out on the ground or in branches. The important thing is that this so-called play behaviour is not as individualistic as it appears. When one studies the matter closely it becomes apparent that the pattern of movements is the same in all members of the species. It must therefore be an innate pattern. The function of the play must be to keep the animal in trim for speedy movement through branches in the day-to-day business of living.

This does not mean that the squirrel may not enjoy its recreation, and one sometimes sees a squirrel travelling through the tree-tops in an apparently pleasurable fashion. Instead of going from one tree to another, taking only short leaps to bridge the gap between outermost twigs of adjacent trees, a squirrel will travel along a line of trees making successive long leaps of 12 ft. (3·6 m.) or so. Then, it sails through the air, merely touching each tree-top to give a push-off for the next leap. This suggests that there is relatively great power in the hind-legs, as in the grey squirrel which made a standing jump from the ground to grasp twigs 10 ft. (3 m.) up. Another in my experience made a standing jump from the ridge tiles of a porch to reach a gutter 8 ft. (2·4 m.) above it.

Considerable strength and dexterity are shown by grey squirrels in taking food put out for birds. In an attempt to defeat them many people hang half a coconut or a wire basket containing peanuts from the middle of a wire stretched between two posts. Squirrels have been seen walking the wire either hanging by all four paws and travelling hand-over-hand, or walking on the wire Blondin-fashion. Over the coconut or basket it will hang by its hind-feet to reach down to the food. This is, however, not remarkable because squirrels often hang like this in trees to reach nuts and berries. I have examined several films showing a squirrel walking a wire, either hanging upside down or on top of the wire. One film showed a squirrel walking under the wire, hanging by its paws, then suddenly, in a lightning movement, swinging up to continue walking on all fours on the wire. There was no preliminary side-to-side swing to help this, and the change from one position to the other suggests remarkable muscle control, possibly also unsuspected strength, and in it the paws must have been used very like hands to give a grip.

The paws can be used, also, to remove lids from tins. One of my investigations concerned a large biscuit tin used to contain bread. I never saw the squirrel at work, but day after day the lid was removed and the bread gnawed; the circumstantial evidence, including paw-marks, pointed to a grey squirrel. Another was seen to turn a smaller biscuit tin upside down and, steadying the top with its front paws, push the lid off with its hind-feet.

Any idea that an animal, even a mammal, may enjoy what it is doing is apt to be frowned upon by those who study animal behaviour. One point against us in arguing this case is that few animals have mobile features like ours to give outward expression. A grey squirrel has a particularly 'dead-pan' face, the eyes especially having a fixed stare. In fact, a squirrel has no need to move its eyes: it can see all round. Moreover, it can detect the slightest movement because its retina is made up of cones only, which increase its sensitiveness to movement.

There is an old legend that when a squirrel swims a river it holds its tail vertically and uses it as a sail. This is not borne out by the observations I have collected. There is plenty of evidence that both red and grey squirrels voluntarily swim across rivers and lakes. A swim of one mile (1·6 km.) has been recorded for the red squirrel and one of five miles (1·6–8 km.), across a lake in North America, for the grey squirrel. By contrast, there have been signs of the spread of grey squirrels in Britain being controlled by rivers. It may be, therefore, as with some other terrestrial species, that a few individuals take far more readily to the water than the rest.

In the water only the head is visible in front, with the body completely submerged; the tail is held out behind but arched, so that the middle of it is above water. Squirrels are described as swimming with a dog-like paddle. On emergence from the water a squirrel is soon dry although the fur is not specially oily. It seems that only the guard hairs get wet and, from other sources (e.g. squirrels seen taking food from under dripping trees) it appears that water is thrown off the guard-hairs by jerking movements of the fore-quarters and head.

There is no record of an unprovoked attack by a wild squirrel although in 1967 a woman reported having been set upon by a grey squirrel when weeding in her garden. She shook the squirrel off, but it came back repeatedly, jumping almost waist high. She assumed she might have been disturbing its hoard, but it is more likely that this was a pet squirrel that had escaped or one that had been made semi-tame by being fed. People have noticed that pet squirrels which have become fixed on one person are liable to attack others, even people with whom they are familiar. The explanation is that wild squirrels form social groups and become pugnacious towards strange squirrels intruding into the group. The aggressiveness of the tame squirrel means that it is treating all but one person as intruders.

It is often said that the grey squirrel ousted the red, but the latter appears to have been decimated by a disease at about the time that the grey squirrel started to spread. There was an instance a few years ago when red squirrels in a given locality developed a mange and died off. Grey squirrels soon moved in to occupy the vacant space. Even so, there is evidence that where the two are in competition the advantages are with the introduced grey. Another general belief is that the red is less destructive and not such a pest as the grey. There is little foundation for this.

Some years ago a campaign for the total extermination of the grey squirrel was undertaken. It is now known that this had little effect, largely due to the grey squirrel's ability to keep out of sight among the trees. In one large garden, for example, it was thought there were a dozen squirrels. As part of the campaign a professional pest exterminator was brought in and he shot thirty-nine squirrels in a few days. Over large areas elsewhere there could not be given this close attention and we can be sure that large numbers of squirrels escaped detection.

Dental formula: i $\frac{1}{1}$, c $\frac{0}{0}$, p $\frac{2}{1}$, m $\frac{3}{3}$ = 22.

The Coypu or Nutria

Family CAPROMYIDAE   *Myocastor coypus*

Coypus are natives of South America; their alternative name, nutria, is Spanish for otter. They were trapped extensively for the unusually fine fur on their underparts, the animals being skinned by cutting along the back. It is one of the largest rodents, being 2½ ft. (76 cm.) long, of which nearly half is tail, and weighing up to 20 lb. (9 kg.). The superficial rat-like appearance (at one time they were exhibited at fairs as 'giant sewer rats') is given by the stout body, squarish muzzle and scaly tail. The hind-feet are webbed.

The farming of coypus in Britain started during the 1930s and animals soon began to escape and colonise rivers and marshes, especially in East Anglia, where they lived on water plants such as reeds, sedges and Canadian waterweed. Until a few years ago they were not regarded as a pest. They helped to clear watercourses by eating the vegetation and they seemed not to burrow into banks. Recently coypus increased in number and their habits appeared to change; they were making burrows through the banks of rivers and drains. Although these were normally above the tide level, an exceptionally high tide would flood through them. Agricultural crops were also attacked, especially sugar beet and kale, and grass and green cereals were grazed.

In 1962 an official campaign to exterminate the coypu started. The rodents had, by now, spread from the Broads of East Anglia to the neighbouring counties and there were occasional colonies farther afield. Trapping and shooting, the best methods of control, were organised systematically by starting in the outlying parts of the range and working inwards. Now, coypus are restricted to the most inaccessible Broads.

Coypus are active at dusk and dawn, but may be seen on frosty days. Breeding occurs throughout the year. Each female has about two litters of two to nine young each year. These are born, after 100 to a 132 days' gestation, in nests of reeds above ground level. They are furred with eyes open at birth and they move around a few hours later. Suckling takes place in the water, the female's teats being high up on the flanks.

Dental formula: $i \frac{1}{1}, \ c \frac{0}{0}, \ p \frac{1}{1}, \ \frac{3}{3} = 20.$

FLESH-EATERS

The Fox

Family CANIDAE   *Vulpes vulpes*

It is usually assumed that, but for its careful preservation by the various 'hunts', the fox would have been extinct long ago except in the wildest and most remote corners of our island. As far back as the reign of Elizabeth I an Act of Parliament was passed for the protection of grain, which incidentally provided for the payment of twelve pence for the head of every fox or gray (badger) that might be brought in to the officers appointed to receive them. Today, outside the hunt areas, the killing of a fox is considered a meritorious act, particularly in the northern mountain districts. Yet in spite of all the fox has survived, due more to its ability to look after itself than to any steps allegedly taken for its preservation. Indeed, at times, in 1965 and 1966 for example, foxes have become unusually numerous, and so bold that even by day they will come quite close to houses or people. Then their numbers drop again, apparently solely from natural causes.

The head and body of the fox measure usually a trifle over 2 ft. (61 cm.) in length, and the bushy white-tipped tail adds up to 16 in. (40 cm.) to its length, but examples have been recorded greatly exceeding these measurements, especially in Scotland. A well-grown fox stands only about 14 in. (35 cm.) at the shoulder. The beautiful fur is sandy, russet or red-brown above and white on the underparts. The backs of the ears are black, as are the fronts of the legs, but these may be brown, and can change from the one to the other with the moult. These might be called the typical colours, but there is a great deal of variation, not only between one individual and another, but in the same individual from one season to another. The foxes (Tods) of Scotland, although of the same species, usually have greyer fur than the English fox. Weights vary considerably, but on average a dog-fox weighs 15 lb. (6·8 kg.), a vixen 12 lb. (5·4 kg.).

(*above*)
Polecat

PLATE 25

(*right*)
Scottish wild
cat

PLATE 26   (*above*) Otter swimming
              (*below*) Portrait of an otter, showing high development of whiskers

The sharp-pointed, long muzzle, the erect ears and the quick movements of the eye with its elliptical pupil, combine to give the fox an alert, cunning appearance, which so impressed the ancient writers that they invented many stories of its astuteness. Dog-fox and vixen are similar in appearance, the vixen being slightly smaller and also distinguished by a narrower face since she lacks the cheek ruffs of the male.

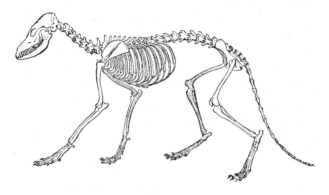

Skeleton of fox

When fully haired the caudal appendage is known as a brush, but in a cub and in adults in full moult it is no more than a tail. The tip (or tag) is white, but may be black. At the moult, in July to August, foxes lose their characteristic appearance. They then look thin-bodied, long-legged and slender of tail, and foxes in full moult have been responsible for at least some of the reports of strange animals seen in the countryside.

Foxes are largely nocturnal, but they can often be seen by day as well. Except at the breeding season dog-fox and vixen lead solitary lives. Most of the day is spent in an 'earth', which is more a cavity in the earth than a burrow. Foxes are said rarely to make the earth by themselves, but to acquire them from badgers or rabbits. In the former instance it is supposed that in taking up quarters in the entrance to a badger's set and rendering it uninhabitable to the more cleanly beast by permeating it with the secretion from glands under the tail, the fox dispossesses the badger. In the case of the rabbit burrow, the fox is said to obtain possession by eating out those who constructed it. These

10

things may be correct. The few earths I have examined seemed to have little to do with badgers or rabbits. On the other hand, foxes in captivity show a strong tendency to dig, and when allowed semi-freedom will construct earths readily by their own efforts. One of the first difficulties in setting forth an account of any of the habits of foxes is that there is little precise information about them on record.

We are told that a fox, having acquired an earth, stops all the exits except one, leaving, where possible, one that opens in a bramble thicket or the dense undergrowth of bracken on a hillside. We could expect the fox to leave the exit that is best hidden, but on the other hand, my experience with foxes in semi-captivity is that they always make at least two openings to the earth. For example, if provided with an earth having only one exit the fox will dig out a second. A great deal has been written about a fox leaving its stronghold at dusk and trotting at a light easy pace along its accustomed trails, keeping a watchful eye for rabbit, hare, pheasant and partridge, or, on winter nights, prowling around farms, looking for a hen-house door which has not been properly secured, or a fowl that is sleeping out in the copse. Certainly foxes will take poultry and they will take lambs, but, as one experienced writer has pointed out, these habits tend to be local. A vixen that has taken to killing poultry will teach her cubs to do the same. Doubtless something of the sort happens with lambs; we cannot be certain, however, as there are few known facts. More solid information comes from a Ministry of Agriculture investigation of the stomach contents of dead foxes. This shows that, now that rabbits are scarce, the chief items of food are rats and mice and also bank voles. These do not represent a full list; hedgehogs, squirrels, voles, frogs, even snails and beetles, as well as a great deal of vegetable matter are also eaten. Birds will be taken, and pheasant, partridge and poultry clearly do fall victim. But not all foxes are habitual poultry stealers and there have been many instances of foxes repeatedly visiting poultry farms or private gardens containing a few poultry and never molesting them.

By contrast, a visiting fox will soon discover offal or carrion, even if buried 2 ft. (61 cm.) in the earth. Foxes also visit garbage bins, and a feature of the many foxes now living in towns is that they turned scavenger. Railway marshalling yards also have their foxes, feeding probably on food thrown out from restaurant cars or on rats coming for this food.

During winter, we can hear at night the characteristic barking, a series of two, three or four short barks, regularly repeated. It is said that when the vixen is in season the dog-foxes gather round her and

Climax to play in foxes

caterwaul, but I have not heard this. Certainly, foxes use a great variety of calls. Once they have mated, the dog and vixen keep close together and there is a great deal of playing between them. In semi-captivity the play consists mainly of 'touch', 'king-of-the-castle' and standing on the hind-legs with the fore-paws on each other's shoulders. This is usually initiated by the dog, and it probably takes place before and after mating. The voice used during play is a throaty staccato barking, somewhat like the call of a jackdaw.

Mating takes place from late December to February. The gestation period is fifty-one to fifty-two days. About April the vixen produces her single litter for the year of, usually, four cubs. They are blind until ten days old, and remain in the earth until nearly a month old, the vixen staying close with them, while the dog plays a large part in supplying the food. When about a month old the cubs come out at night for their first exercise. In fox-hunting counties artificial burrows are constructed in suitable places, of earth and stone, of which the expectant-mother vixen will avail herself. These are furnished in order that the cubs may be dug out with ease when they have reached the proper age for the huntsman's purpose.

The evidence from families in captivity suggests that the departure of the cubs is determined to some extent by an increasing irritability of the vixen, when the cubs are about two months old. The cubs reach adult size six months after birth, and become sexually mature in their first winter.

Soon after the cubs come out of the earth for the first time they can be seen playing as a group with the parents outside the earth, and this continues for several weeks. Later, the vixen takes them hunting at night, so they learn from her example how to fend for themselves. In due course, the cubs tend to wander and this, combined with the vixen's increasing irritability, ensures that ultimately they do not return. Probably the young males go first. The dog brings food to the vixen when she is lying in the earth, waiting for her cubs to be born. When doing this he uses a particular call to bring her out to take the food from his mouth. He continues this after the cubs are born. In semi-captivity the dog-fox continues to do this even after the cubs are weaned. Then the cubs take the food from his mouth themselves, or the vixen may take it and the cubs take it from her mouth. The cubs have to jump up to reach the parent's mouth, and all the time the parent is moving its head, from side to side or up and down. There can be little doubt that the cubs having to work for each morsel are not only being exercised, so developing their limbs, but are learning to co-ordinate movements and senses. During the time, also, the dog plays a great deal with them, more so than does the vixen, who sits sedately by, merely watching.

Foxes are credited with resorting to a particular stratagem to attain their end. A story is usually told of a fox which, seeing a party of rabbits feeding, and knowing that they will bolt to their holes on its approach, starts rolling about at a safe distance to attract their attention. Then like a kitten it will begin chasing its tail, while the silly rabbits gaze, spellbound, at the performance. The fox continues without a pause, as though oblivious to the presence of spectators, but all the time it is contriving to get nearer, until a sudden straightening of the body enables it to grab the nearest rabbit in its jaws. Certainly, something of this kind does take place, and it has been called 'charming'.

There are too many authentic accounts of foxes charming to leave much doubt in the matter. From these, a more likely explanation evolves: foxes are naturally playful. Like some other mammals they will, without obvious cause, suddenly behave as if they have taken leave of their senses, bounding about, bucking, somersaulting, and so on. Rabbits and birds on seeing these antics are drawn to watch out of curiosity. If the fox is hungry then the spectators suffer. It is possible that a fox playing in this way, and finding birds and rabbits attracted to it, might use it again, with more deliberation. Such learning by experience would not be beyond a fox's intelligence, but there is much to be said for the view that charming, as such, is not primarily a deliberate stratagem. At all events, there is no question that foxes will play, on their own or with other foxes. There is even a well authenticated instance of a fox playing with a leveret and, at the end of the play, departing and leaving the leveret unharmed.

Although the red fox lives mainly on the ground there are numerous instances of it climbing trees. Usually this occurs when a tree is leaning or when there is a trailing bough that has broken and is hanging down to the ground, up which the fox can clamber. There are many known cases of foxes taking abode in the tops of pollarded willows, and not infrequently they have been seen well up in elms, to a height of 30 ft. (9 m.) or so. Perhaps the most remarkable single instance is of a fox that had its sleeping place at the top of a bole of an elm, which had been previously pollarded but now had several stout branches rising from the top of the bole. The fox's nest was 14 ft. (4·2 m.) from the ground, with no branches between it and the ground. Nobody had seen the fox climb the tree although several had seen it come down. Scratches on the bark showed that the fox jumped on to a large buttress root and then scrambled up the bole to the nest.

The red fox ranges over Europe and over Asia as far south as central India. It is found throughout the British Isles, except for Orkney and Shetland and all Scottish islands other than Skye. In central Asia it lives up to 14,000 ft. (4,268 m.) above sea-level, and in Britain is found at all levels. The foxes of the northern hill country are said to be a finer race than those of the southern woodlands. This may be because every man's hand is against them, and because only individuals of great resource and superior physique survive to continue their kind. It may be, on

Teeth of fox

the other hand, the result of climatic conditions, more especially of the generally lower temperature.

Dental formula: i $\frac{3}{3}$, c $\frac{1}{1}$, p $\frac{4}{4}$, m $\frac{2}{3}$ = 42.

The Feral Dog

Family CANIDAE   *Canis familiaris*

Since August 1964, there have been over three hundred reports of people claiming to have seen a puma at large in various parts of the southern half of England, from Cornwall to Norfolk. Except for the now-famous Surrey puma, all the rest were positively identified as feral dogs or feral cats, and there is reason to believe that several dogs and at least one feral cat formed the backbone of the Surrey puma story. This served to focus attention on the subject and it soon became clear that both feral dogs and feral cats are as much a part of the fauna of Britain as feral mink or introduced deer, and that they have been with us much longer than these two. Although information about them is extremely hard to obtain it seems worth-while setting this on record.

The puma story originated with imperfect sightings of an unusually large ginger feral cat, near Farnham, Surrey, but was firmly established when large paw-marks $5\frac{1}{4}$ in. (139 mm.) across were found on a sand track used for exercising racehorses, at Munstead, to the south of Farnham. These paw-marks were identified at the London Zoo as 'puma'. The following year I identified almost identical tracks on Hurtwood Common, to the east of Munstead, as those of a large dog bloodhound. Some zoologists expressed scepticism that any dog could leave such large tracks, but this is put beyond dispute by photographs and plaster casts I took using this individual bloodhound.

Several times during the period of the 'puma scare' other large tracks were identified by London Zoo and Bristol Zoo as 'puma' which proved later to have been made by hounds (e.g. bloodhound, great dane). All were 5–$5\frac{1}{2}$ in. (126–138 mm.) across. Even a bassett hound may leave a paw-mark (from the front paws) up to $4\frac{1}{2}$ in. (114 mm.) across. These large hound tracks can be even more misleading because they often show no claw-marks.

Feral dogs tend to be of the larger breeds, such as alsatian, labrador and greyhound types. From film and photographic records of them certain characteristics emerge. They tend to carry their hind-quarters lower than the fore-quarters, which gives them a gait suggestive of a hyaena. Their run is somewhat 'bouncy', and this is true of others beside the greyhound type, which normally has a bouncing action. The feature which, above all, marks them off from the truly domestic dog is that they turn the head constantly from side to side while resting crouched; to some extent they do this while running. This last feature, presumably, is linked with the absence of security, and is the method of keeping watch for possible intruders.

The distinction must be made between dogs that stray for days or possibly weeks on end but are later retrieved, and those that have 'gone wild'. One greyhound is known to have lived for three years in an area lying between Farnham and Guildford, in Surrey. It was sufficiently tame to approach human dwellings to search for food, yet rapidly made off if approached. Its sleeping place in a wood was marked by the flattened grass.

Another feral dog, of the labrador type, had its home range several miles to the east of the greyhound. Like it, the labrador was rarely seen, would approach houses, allow children to go up to it, but made off the moment an adult approached. It would occupy one sleeping place for days on end, then move on.

These two are typical of many feral dogs brought to my notice, and there are probably many more living wild but whose presence remains unsuspected. For example, a wood in Kent was drawn because the foxes in it were too numerous. The beaters flushed a feral dog that must have been living there for several years. It, apparently, had several resting places, all marked by an area, several feet across, where the vegetation was flattened.

Dental formula: i $\frac{3}{3}$, c $\frac{1}{1}$, p $\frac{4}{4}$, $\frac{2}{3}$ = 42.

The Badger

Family MUSTELIDAE *Meles meles*

The badger has been described as the oldest inhabitant of Britain, a romantic allusion to its survival from pre-historic times. The same thing can be said of several other of our wild animals, but whereas they have been seriously reduced in numbers, or wiped out over much of their former ranges, by human persecution, badgers have persisted in spite of it. Now, with an increasing benevolence shown to them, they are in much less danger. Badgers are present in every county of England, Wales and Ireland, and in most counties of Scotland although not very common there, but they are not found on the islands around Britain, with the exception of Anglesey.

The tenacity in survival shown by the badger is in part due to its habits. It is, with the hedgehog, our most completely nocturnal animal.

It is sometimes seen abroad during the day, but on the whole it comes out regularly after sunset and goes home again at dawn. Its habit of living deep in the earth is in its favour, and above all it keeps very much out of the way, warned by acute senses linked with a wary disposition. There must be many people who have lived near a family of badgers without suspecting their presence. Badgers are even living in the parks of outer London, unbeknown to the majority of visitors to the parks.

The badger measures 2½–3 ft. (75–93 cm.) long and stands about 1 ft. (30 cm.) high at the shoulder. At a short distance its rough coat

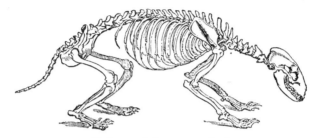

Skeleton of the common badger

appears a uniform grey, but on closer inspection it is seen to be some-what reddish on the back and black underneath. There is something almost pig-like about the stout, squat body, yet its gait is more reminis-cent of that of a bear. The legs are short, there are five toes on each foot and the front paws, more especially, are armed with powerful claws. For its size it is immensely strong, and very active when awake, but it has the gift of sleeping soundly in an attitude of complete repose.

The most striking feature is the head, with its long tapering muzzle marked with broad bands of white, and with a black band running from the ear to the snout, on each side of the head. The whole appearance of the head suggests that an acute sense of smell predominates over the other senses, for the ears and eyes are both small; this is in fact true. That does not mean that it cannot hear well, for the slightest sound will set the badger on the alert, but from the way it behaves it is fairly clear that the eyes rank a poor third after smell and hearing. Whereas in most animals the lower jaw falls away from the skull when the flesh is removed, in the Mustelidae it is hinged and cannot be removed without breaking

the bony socket in which it articulates, so there can be no sideways movement of the jaw and it cannot be dislocated without fracturing the skull.
This accounts for a badger's tenacious grip.

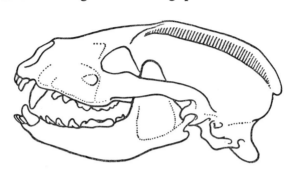

Skull of badger showing prominent, bony ridge

Its diet is fairly wide, and although classified with the Carnivora, and
having large canine teeth, it is essentially omnivorous. It will take acorns,
beech mast and other wild fruits, eat grass and clover and will dig for
succulent roots. It will search for beetles among the dead leaves and
decaying wood, feeding especially on the larvae of beetles and of night-
flying moths, on earthworms and slugs, and will dig out the nests of
bumble bees and wasps for the grubs. Young rabbits are dug out, and
it appears that other forms of flesh are also taken, such as lizards, mice,
voles and young birds fallen from the nest, as well as still-born lambs.
Badgers have also been known to maul carcases of deer.

The badger's underground home, known as a set, may be 10 ft. (3 m.)
or more below the surface, with a main entrance sloping down to the
passages and upper and lower galleries, with probably a back door at
some distance from the main entrance. The main opening is marked
by a mound of earth, turned out during the excavation, and the size of
this gives some indication of the depth and extent of the set as a whole.
A special breeding chamber, furnished with moss and grass, is used for
the reception of the cubs. These are usually born in February, although
litters as early as mid January and as late as May have been known. The
normal litter is two to three, but there may be one cub only, or as many
as five. Sometimes two families may be seen together, giving the impres-

sion of a large litter. Under normal conditions the cubs remain below ground for six to eight weeks after birth.

The new-born cub is just under 5 in. (120 mm.) long, blind and a dirty white in colour, with no hair on the undersides. The eyes open at about ten days. At first the head stripes are faint, but about the time the eyes open the hair is becoming darker, a light grey, and at the same time the dark lines on the head become more emphasised. For a time, after first coming above ground, the cubs do not wander far from the entrance to the set. The mother comes out first and, if the coast is clear, will turn back into the entrance, possibly calling the cubs out. They may keep close to her, even to the point of crawling under her. At this time, too, they will bolt into the entrance at the slightest disturbance; if they do not, and there is real cause of alarm, the mother will yelp and push them down in front of her.

At about eleven weeks the cubs become more active, often being the first to appear at the entrance to the set. They are then beginning to play more, but although becoming more independent they are still very wary and readily return underground at the slightest disturbance. They are weaned a week or so after this, but although they soon begin to forage for themselves they remain with the sow until autumn, sometimes throughout the winter. The females become sexually mature at a year to fifteen months of age; the males at one to two years.

Although the cubs are born in February, mating usually takes place in July but can occur at any time between February and October. This does not mean a gestation period of seven months, because there is a delayed implantation. That is, the fertilised ovum remains free in the oviduct, develops up to the blastocyst stage and then development is arrested. Seven to eight weeks before the actual date of birth the embryo becomes implanted in the wall of the uterus and development proceeds normally.

During the mating season, the boar and sow indulge in a good deal of play, and at such times the air may be heavy with the scent of musk. The cubs of the year are then still with the parents, and do not leave them until September or later. They do not themselves breed until the following July or later, after which the female will have one litter a year, not every other year as sometimes stated.

The weight of an adult male badger is about 27 lb. (12·2 kg.), the female being a few pounds lighter, but some males have weighed up to 40 lb. (18 kg.) or a little more. There are three records of weights of 60 lb. (27 kg.) or over.

Badgers in this country do not hibernate, although they may be less active in the winter. They are said to hibernate partially in the colder parts of Europe, and even here they are known to be able to go for as much as fourteen weeks without food, living on fat stored under the skin. It is possible that badgers may have been hibernants originally, but have lost the habit as the weather has become milder. Fossil remains show the badger to have been in existence for at least 250,000 years.

Another point on which misleading comment is often made relates to the white markings on the face. It is said that these are protective because they harmonise with the light and shadow of a moonlit night. Badgers avoid moonlit nights by keeping underground, or using dense cover. More likely these are recognition marks by which the animals can see each other, in the darkness of the set or above ground. Whether they are warning signs to enemies, as has been suggested, must remain a matter of opinion.

Badgers are proverbially clean animals. They use latrines dug in the ground at a distance from the set, or they may have special chambers in the set for these. In January and February the set is spring-cleaned, old bedding and earth being brought out and the bedding renewed, both sexes sharing the work. The main time for taking in new bedding is in autumn, but it can take place in almost any month. A badger collects bracken, dried leaves or other vegetation and, with this held between the chin and fore-paws, it shuffles backwards down the entrance to the set.

A small cub makes a high-pitched whickering, and a loud squeal when alarmed. Cubs playing make puppy-like noises. Adults growl or bark as a warning and purr with pleasure, and both boar and sow will scream, a long-drawn eerie sound, often for no obvious reason.

Badgers are said to bury their dead.

The minute first pre-molar in each jaw is frequently shed early, and may be missing from any adult skull examined. The badger's dental formula is: i $\frac{3}{3}$, c $\frac{1}{1}$, p $\frac{4}{4}$, m $\frac{1}{2}$ = 38.

The Otter

Family MUSTELIDAE   *Lutra lutra*

Otters are more common in this country than is usually imagined. One may meet one accidentally, not uncommonly in the headlights of a vehicle, or by deliberate search, on the banks of remote streams or tarns, where there are alder-holts, or in the neighbourhood of the East Anglian Broads. It may sometimes be found by day in summer, lying up in the caves on some remote part of the coast where the cliffs are rocky and the shore strewn with boulders. The occasional otter has sometimes been seen in the Thames, in the neighbourhood of the heavily built-up areas of London.

The long, lithe body of the otter, ending in a long tapering tail, thick at the base, is streamlined for swimming. The head is broad and flattened from above, the face short, the black eyes small but bright. The short, rounded ears are hairy and do not project beyond the fur. The legs are short and powerful, and all feet are completely webbed. There are five toes on each foot, those on the fore-feet having short pointed claws. The claws of the hind-feet are more flattened and nail-like. The tail is somewhat flattened from the sides, and is used as a rudder. Below its thick base there is a pair of glands which secrete a fetid fluid. The fur is of two kinds; a fine soft, waterproof underfur of whitish-grey with brown tips, interspersed with longer and thicker glossy hairs, the so-called guard-hairs. On the upper parts and the outer sides of the limbs, these longer hairs, which have a grey base, have rich brown ends, so that the whole body appears a rich brown. The hairs on the cheeks, throat and underparts are light brown to silvery grey. There is probably an autumn moult, but little is known about it. One suggestion is that there is only one moult which takes place over a large part of the year and progresses almost imperceptibly.

When swimming under water, where much of the hunting is carried out, the ears and nostrils are closed. In clear water the eyes could be used for following the quarry, but in turbulent or muddy water the whiskers are almost certainly used. These are stout and set in a glandular moustachial pad which is richly supplied with nerves, a sure sign that it is highly sensitive. Doubtless the whiskers pick up vibrations in the water and thus guide the otter.

As soon as the otter submerges, the guard-hairs become wetted and lie flat over the underfur, largely preventing water from entering it. In addition, the dense underfur carries air trapped among the hairs, so that any water passing the guard-hairs is excluded. The trapped air also serves to insulate the body, preventing loss of heat. On climbing out on to the bank, water runs off the guard-hairs, causing them to bunch together in groups tapering at the tip, giving a spined effect to the coat. A vigorous shake by the otter throws the water off the guard-hairs and the coat soon dries and resumes its normal appearance.

Skeleton of the otter

The total length is about 4 ft. (1·2 m.), of which ⅓ is tail, but there are records, from skins, of a length of 5 ft. 3 in. (1·6 m.). The weight of a full-grown male is 20–25 lb. (9–12 kg.), but occasionally it exceeds 27 lb. (12·2 kg.). There are records of 50 lb. (23 kg.) and one of almost 60 lb. (27 kg.), and in the sporting books of fifty or more years ago are a number of references to otters seen that looked startlingly big.

Its 'holt' or lair will probably be a hole in the bank with the entrance under water and overhung by alders and rank herbage, but it may be well away from water. The holt is used only for the time that cubs are on hand; at other times an otter has no permanent home. There may be an alternative way into the holt at the back of the bank above water. A short distance in from the mouth of the tunnel, a side-chamber will be found, which is the family midden. The otter rests in the day-time in the holt or in reed beds, coiled up like a dog with its tail around its face. The 'spraints', or droppings, are a good clue to the otter's presence. These are black when fresh, turning paler later.

Otters start their hunting about sunset, when their flute-like whistle can be heard. Other vocalisations are the whickering of the cubs, a long drawn out moan, a low pitched chortle of pleasure and hiss or high-pitched chatter of annoyance. Essentially a fisher, it always brings its catch to the bank to be consumed. The backbone of a fish is first bitten through behind the gills; and where the fish are large and plentiful an otter may take no more than a bite from the shoulder. At other times it may eat methodically from this point to the tail, which is always left. Except that it must frequently surface to breathe, it is as much at home in the water as a fish, swimming in circles where the water is deep, its movements as graceful as those of the fish it pursues. An otter is often a wasteful feeder, killing more than it needs, but there is reason to believe that its victims are often weak or sickly fish. An important item in its diet is eels, the bane of the angler. Occasionally an otter may take wild ducks or moorhens and on land, rabbits, rats, mice and voles. Frogs, newts and freshwater shrimps are also eaten, and on land slugs, earthworms and beetles. A favourite food is crayfish. A characteristic trick is to float with the current downstream, the fore-legs pressed against the sides and only the upper part of the head with eyes, ears and nostrils exposed above water.

In summer, when water is low in the stream, an otter may travel across country, from pool to pool by night, to an estuary or to the open coast. Although so obviously adapted for an aquatic life, an otter can travel with speed on land and will cover about 15 miles (24 km.) in a night. On the coast it will use a cave as a shelter, from which it will work the shallow waters for flatfish, bass, crabs and mussels. With the autumn it will return to its usual inland haunts, perhaps taking the migrating eels on the way.

Otters hunting the coasts are sometimes spoken of as sea-otters, a name which should be reserved for a totally different species living in the north-east Pacific.

Relatively little is known about the breeding habits. There seems to be no fixed breeding season, the young being born at all times of the year. The gestation period is said to be sixty-three days, but there may be delayed implantation and there are two to three young in a litter. A nursery nest or 'hover' is constructed of rushes and grass, lined with the soft purple flower panicles of the great reed. In Norfolk the nursery is frequently found on the surface of the water, in the great reed-beds. The cubs are born blind, but already covered with a fine downy fur. The eyes open at thirty-five days. Both parents hunt to provide them with food, and they remain in the hover for several weeks before being taken out to be taught their way about in the water. It is believed that the partnership of the parents is only temporary, and that as soon as the young ones are capable of taking care of themselves the dog otter goes to live by himself. The cubs probably remain with the mother until she is ready to breed again.

Otters, like other carnivores, are given to playing. One of their particular games is tobogganing down muddy slopes, and especially over snow-covered or icy ground. Even in their normal progression overland otters often take two bounds and a slide, two bounds and a slide.

Widespread in Britain the otter ranges across Europe and North Africa.

One trick of otter behaviour that has been almost entirely overlooked is their habit of occasionally swimming in line astern. This is seen mainly around dawn. The bitch and her cubs will do this and on rare occasions several such family parties will follow each other through the water, showing nine to twelve humps with the extended neck and small head of the leading otter showing above water. The appearance then is

Eye witness drawing of appearance of an otter bitch followed by her cubs

virtually identical with that of the traditional picture of the sea serpent, and there can be little doubt that many stories of lake monsters, throughout the northern hemisphere and in Africa, are based on this. While swimming in this way the otters are evidently very alert, and the slightest sound causes all to submerge simultaneously in a fraction of a second. From the evidence I have gathered it seems certain that this behaviour is not indulged in except in undisturbed lakes.

PLATE 27 (*above*) Stoat in snow
(*below*) Weasel and its prey, a wood mouse, to show relative size

(*above*) Red deer stag roaring

PLATE 28

(*left*) Sika deer

Otter have given rise to more stories of mystery animals on land and monsters in water than any other species. Apart from the swimming in line astern a pair of otters will leap at the surface, one behind the other, in pursuit of a large fish. The effect is of a large fast-moving writhing snake. Another trick is to swim with only the head exposed, paddling rapidly with the hind-feet, leaving a line of dark hump-like waves in the almost unseen animal's wake.

The lower jaw is hinged to the skull, as in the badger, and can move only up and down. The dental formula is: i $\frac{3}{3}$, c $\frac{1}{1}$, p $\frac{4}{3}$, m $\frac{1}{2}=36$. The molar teeth have sharp tubercules on the crown.

The Pine Marten

Family MUSTELIDAE   *Martes martes*

The pine marten is now found only in the wilder parts of the Lake District, the north of England, Wales, Scotland and Ireland, where it exists in small numbers. In some of these places, notably in the Lake District, its numbers are tending to increase.

The pine marten, or marten cat, was formerly a common woodland animal, and in the Middle Ages 'hunting the mart' was almost a national pastime. It was hunted by groups of men on foot with sticks and stones, but the real reduction of its numbers came in the nineteenth century, with the onslaughts of the gamekeeper and the high prices paid for the skins. In addition, the use of the gin-trap for rabbits resulted in something little short of a massacre of the pine marten. The last killed in the London region was shot in Epping Forest in 1883.

The pine marten resembles the better-known stoat but is larger, with relatively longer legs, a broader, more triangular head with sharp-pointed muzzle, and a longer, more bushy tail. Its entire length is between 25–30 in. (63–76 cm.), of which 9–12 in. (25–28 cm.) is tail; the females are slightly smaller than the males. The weight of adults ranges from about 2–3⅓ lb. (0·9–1·5 kg.). Its coat is a rich warm brown, except on the throat and breast where the colour varies from orange through yellow to creamy-white. The middle of the back and the exposed sides of the legs and feet are darker than the rest of the body, and the underparts are greyish. The superficial colour is provided by the long upper glossy fur, beneath which is a finer and softer underfur of short reddish-grey hairs tipped with yellow. One moult lasts from late spring to June, starting on head and legs, to give the short summer coat. A winter moult starts in September or October, beginning on tail and legs. The eyes are large, black and prominent, the ears broad, open and rounded at the tips. Like all the members of the family Mustelidae, the marten has special scent glands near the base of the tail. It is these which enable the skunk and the polecat to disgust their enemies, but in the marten the secretion has a musky odour and is not objectionable; in consequence one of its old English names was sweet marten to distinguish it from the foulmart or polecat.

The pine marten is mainly arboreal, although in Sutherland its chief habitat is open rocky ground. The long, slender body and sharp claws specially fit it for climbing, the long bushy tail being used as a balancer in negotiating slender branches in the pursuit of birds, or in reaching their nests for eggs. A marten is at least as agile in trees as a squirrel, being able to climb rapidly, and to negotiate even slender branches that give under its weight. It can also leap across wide distances to land with precision on a small foothold. Indeed, the pine marten is one of the most persistent enemies of squirrels, and in destroying it the countryside was laid open to the almost unrestricted spread of the grey squirrel. It is by no means unknown for a female marten to kill a squirrel, and take over its nest, relining it to suit her own purposes.

Although so much at home in trees, the marten is at times very active on the ground where it destroys rats, mice, and especially short-tailed voles, as well as rabbits, hares and, if the opportunity arises, game-birds and domestic poultry, large and small. It has been accused of attacking lambs and stealing trout from fishing boats. It also eats caterpillars, beetles and carrion. Although a typical carnivore, the marten has a taste for bilberries, strawberries, cherries, raspberries and blackberries. It also robs beehives of their honey.

Mating is in July and August, but there is a delayed implantation, until the following January. The female makes a nest of grass among rocks, in a hollow tree or in an old crow's nest which she relines. She produces a single litter each year in late March or April; the litters vary in size from two to seven, with four or five young on average. These are weaned at six to seven months and leave the nest soon after. The young are easily tamed, but a captured adult is savage and untamable. This is consistent with the natural ferocity and courage of the animal. An adult marten will not hesitate to attack opponents larger than itself, and in an encounter even the wild cat is likely to be beaten.

Martens are usually silent but use a high-pitched chattering in aggression and a deep huff as an alarm note. Mating is accompanied by purring and growling.

An older spelling of the popular name was martin, although originally it was martern, but in more recent works, to avoid confusion with the birds of that name, there has been a general agreement to use *e* as the second vowel.

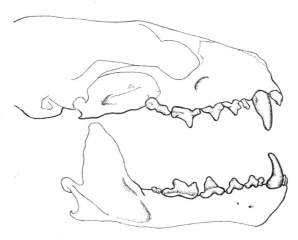

Teeth of pine marten

The pine marten is found in all the wooded regions of Europe and in parts of Asia, northwards from the Mediterranean to the limits of tree-growth.

The dentition of the marten is: i $\frac{3}{3}$, c $\frac{1}{1}$, p $\frac{4}{4}$, m $\frac{1}{2}$ = 38.

The Stoat

Family MUSTELIDAE   *Mustela erminea*

The gun and the snare of the gamekeeper and the poultry-farmer have levied their toll on the stoat, as on the polecat, and the keeper's gibbet always showed a row of stoats. Nevertheless, the species managed to keep itself well represented, even in the strictly preserved woods of southern England, although today it is not commonly seen.

The stoat is up to 17 in. (440 mm.) or more in length, of which about 4½ in. (114 mm.) are the long-haired tail. Males are slightly larger than females and weigh 7 oz.–1 lb. (200–440 gm.), females weighing 5–10 oz. (140–280 gm.). The upper parts are reddish-brown, and the underparts white tinged with yellow. The tail is the same colour as the back, except that the tip is invariably a tuft of long black hairs. Like the polecat, the stoat can secrete an objectionable odour from its scent glands, but this is not quite so offensive as in the larger relative.

In the mountainous districts of Scotland, as in the other northern countries of Europe, the fur in winter becomes pure white all over, with the exception of the tip of the tail, which always remains black. It is then known as ermine. This change of coat also takes place in the north of England, but not so commonly, and in the south it occurs rarely, is often incomplete, and where one white or partially white stoat is seen there are likely to be others in that same locality. In partly white stoats brown patches are confined mainly to the head and back. Sometimes they may be no more than a ring around each eye, producing a spectacled appearance. The accepted idea is that the summer coat is 'protective' in that it harmonises generally with the colour of the leaf litter over which the stoat is moving, and that on snow-covered ground, as in Alpine districts, the brown fur would render the animal conspicuous so that it would be heavily handicapped in its hunt for food. With the change to a white fur, however, the stoat is enabled to steal up on its prey unseen, even from a short distance. The change is fairly sudden, and was formerly supposed to be linked with an appropriate fall in the temperature. More will be said on this later.

A stoat hunts largely by scent, picking up the trail of its prey and following this relentlessly. It moves characteristically in a succession of low bounds in which the long, lithe body assumes almost a snake-like appearance. It can swim and climb well. Although the sense of smell is acute and plays so large a part in the chase, hearing is also acute, but its sight is not. Whether hunting or not, the stoat is alert, agile and energetic, with a natural ability to take advantage of cover, in which it will disappear only to reappear shortly in another spot. A common trick is to use the runs of moles or rats, either to escape enemies or to hunt prey. We have an indication of the effect of its manoeuvres on its living prey from the way a rabbit will cry in terror, apparently paralysed with fear, while the stoat is yet some way off. Under these circumstances a wild rabbit has been known to approach a group of men, apparently seeking refuge and protection, even allowing itself to be picked up, its natural fear of man being not nearly so great as the terror inspired by the stoat. It is said that a hare, under similar conditions, will not exert itself greatly to escape the stoat, but becomes so terrorised as to be unable to adopt methods with which it might outwit a fox or a pack of trained hounds.

Stoats also use charming (see p. 123).

Stoats, truly carnivorous, reject little that is flesh. They hunt along the hedgerows, across fields, by rivers and brooks or wherever there is a chance of food. All members of the weasel family readily take fish, and an eel or other fish placed in a trap is fairly certain bait for them. Where rabbits are common these are a frequent prey. Given the opportunity a stoat can be destructive to game and poultry. The fact that the animal also destroys what is, in an agricultural sense, vermin is not so commonly stressed. It will, for example, destroy moles, as well as rats, although its prey is more commonly the smaller rodents, such as mice and voles, killed by biting the back of the neck. It also takes small birds, eggs and reptiles.

Although largely nocturnal in its habits, it is not exclusively so, and there is more chance of seeing a stoat hunting in broad daylight than of seeing most of our native carnivores.

When the kittens are growing up, parents and young hunt in a family party, and it may be that, like some of the larger carnivores, two or more family parties join up to give the well-known stories of packs of stoats, which are said to attack a dog or even a man. It is easy to believe that these fearless animals, especially with the confidence born of numbers, could be formidable. There are also the occasions when, through increase in their numbers, the food supply of a district has been largely reduced and stoats will migrate in large numbers. There are reports of several scores of them moving across country in a column. These occasions are probably rare and therefore seen only by the chance observer. As a result such stories are received with caution by many zoologists.

The nursery is made in a hole in a bank or the hollow of a decayed tree, the female, who is slightly smaller than the male, alone tending the young. Fertile matings take place in March and again in June and July, and because the males are partially sexually active until October matings that are infertile may take place after July. After fertilisation in spring and summer implantation is delayed until the following spring, after which there is a gestation period of twenty to twenty-eight days. In April or May the stoat gives birth to four or five young, more often six to nine, which she will defend fiercely against all dangers. She has only one litter a year. Weaning is at five weeks of age. The babies have fine white hair covering the body at birth, the black tip appears on the tail at twenty days and the eyes open at twenty-seven days. They remain with the mother after weaning, but the young females soon become sexually mature and mate with older males.

The distribution of the stoat extends eastwards from Great Britain into Asia, and from the Alps and Pyrenees across Europe to its Arctic shores. A local race, smaller in size, and varying somewhat in colour, and known as *Mustela erminea hibernica*, is found in Ireland and is known as the weasel, but no specimens of skins of the true weasel (*M. nivalis*) have ever been received from that country. Another local race, the Islay stoat (*M. erminea ricinae*), is found on the Isle of Jura, on the west coast of Scotland. The same species is in North America, where it is called the short-tailed weasel, from about forty degrees latitude northwards.

The idea, mentioned earlier, of the colour of the stoat's pelage being protective receives support from the fact that where snow is normal in winter it changes to white. Yet it is not easy to reconcile this notion that the colour of the coat permits the animal to creep up unobserved on its prey when we recall the behaviour of rabbits and hares in the presence of a stoat. Their terror must almost certainly be induced by the sight of the animal. Even if it is induced by smell, the argument is not invalidated. It still suggests that any camouflage is not of primary value.

It was believed for a long time that the change in the colour of the coat in autumn was caused not by the loss of hairs but by the loss of pigment in the hairs. This, it was supposed, was carried out by phagocytes, or wandering cells, migrating into the hairs and ingesting the pigment granules. This is now known to be incorrect, at least in the case of stoats and weasels. Although the process has not been fully investigated, there is good reason for believing that the change in colour of their coat is partly dependent upon temperature (and the temperature of the previous winter at that) and partly upon the length of day. However, it must be admitted that the causes are not completely known.

One feature of the change which suggested that it might be due to phagocytes was the rapidity with which it took place. In fact, this is because the new white coat grows underneath the old one before that is moulted in autumn. So, when it is finally shed the newly grown white coat is disclosed, and the shedding may take as little as three days, although under mild conditions it may occupy as much as three weeks. In experimental tests most of the stoats turned white if exposed to cold before or during the moult, but it is significant that not all of them underwent the change. In northern latitudes, on the other hand, all wild stoats turn white, and there we have the annual combination of lowered temperature and the lengthened hours of daylight. One of the indications that temperature alone is not the cause of the change is that, although it may affect the time of the spring moult, it has no influence on the autumn moult. The latter takes place regularly in November to December, whereas the time of the spring moult varies according to latitude. A possible influence of daylight is also suggested by experiments carried out on American weasels. These were subjected to temperatures of not less than 50°F (10°C), but they were kept in artificial illumination, the equivalent of a lengthened day, and this was sufficient to induce the colour change.

There are other anomalies in this matter. Thus, in the south of England, an occasional stoat will turn white in the autumn, or it may be partly white and partly coloured. Both may occur in a mild winter when there is a complete or almost complete absence of snow. One explanation for this is that temperature may also have a delayed action, so that a stoat experiencing lowered temperatures in one autumn may turn white in the following autumn, even if temperatures are high. But the fact that white or partly white stoats, in southern England, tend to be localised suggests that the change to white may be due to genetic causes.

Stoats have few enemies apart from man. Young stoats, and a few adults, are taken by owls and hawks.

Dental formula: i $\frac{3}{3}$, c $\frac{1}{1}$, p $\frac{3}{3}$, m $\frac{1}{2}$ = 34.

The Weasel

Family MUSTELIDAE   *Mustela nivalis*

Although of similar form to the stoat, the weasel is smaller and lacks the black tip to the tail. In general colouring, however, there is little difference between the two, except that the upper parts of the weasel are a redder brown and the underparts a purer white, and the line of demarcation between the colours is less pronounced than in the stoat.

Skeleton of weasel

The head is narrower and the legs are shorter, and the tail less bushy and little more than half the length of the stoat's tail. The average length of the adult male is 8½ in. (200 mm.), with 2¾ in. (60 mm.) of tail. The female is 1 in. (25 mm.) or so less in total length than the male, and weighs, on average, 2 oz. (59 gm.) against his 4 oz. (115 gm.). Because

PLATE 29　(*above*) Day-old roe kid　(*below*) Fallow buck in velvet

PLATE 30 (*above, left*) Pine marten (*above, right*) American mink
(*below*) Coypu

PLATE 31   (*above*) Young roe buck   (*below*) Red stags in velvet

(*left*) Sand lizard on ling, on a Surrey heath

PLATE 32

(*below*) Slow-worms

of this smaller size, the females are known in some districts of England
as cane weasels, under the impression that they represent a distinct
species. There has also been the suggestion, which may have been
influenced by the small size of the females, that we have another species
in Britain, the least weasel (*Mustela rixosa*), but although this species
extends from western Europe to eastern Asia and North America,
it has not so far been found in this country in spite of close searches
for it.

The long slender body, short limbs, long neck and small head give
the weasel a snake-like appearance which is heightened by its active
gliding movements. Like other members of its family the weasel is
courageous and ferocious out of proportion to its size, and will attack
animals larger than itself. It has been seen, in the neighbourhood of a
barn, struggling to haul along a nearly full-grown rat, two or three times
its own weight, killed by a bite through the base of the skull. Sometimes
weasels hunt in pairs or in family parties. There is a saying that a weasel
can pass through a wedding ring. It is hard to persuade a living weasel
to perform this trick, or to go through a hole of similar diameter, but
at least we can say that a weasel's skull can be passed through a wedding
ring.

Although the weasel is chiefly nocturnal, it may also be active by day.
Its diet includes rats, mice, voles, moles, frogs, small birds, eggs, rarely
carrion, and occasionally it has killed poultry. It may swim in pursuit
of the water vole, or climb trees and bushes to rob birds' nests of eggs
or young. Voles and mice are, however, the principal victims, a weasel's
small size enabling it to pursue these rodents in their underground
runs.

The normal method of hunting is to stalk or trail the prey, but
'charming' is sometimes used (see also p. 123). In this the weasel throws
itself into all manner of contortions. Curiosity overcomes the natural
fears of small birds in the vicinity and they approach nearer and nearer
to the performing weasel. When one of them is close enough it is seized
by the 'charmer'. The sudden onslaught may cause the birds to retreat
rapidly, or they may, with the confidence of numbers, fly at the weasel,
mobbing and pecking it, driving it into cover. The weasel's chief
enemies, apart from man, are hawks and owls.

As do stoats, in more northerly latitudes weasels undergo a change of coat in autumn. As a rule, however, there is no seasonal change of colour in their fur in this country, although the occasional individual may be white or partially white in winter. The causes of the change appear to be the same as in stoats (q.v.).

The female weasel builds her nest in a hole in a bank or low down in a hollow tree. It consists of dry leaves, grass, moss and the like. Four to six—usually five—kittens are born in spring or early summer. Pregnancy may occur in any month from March to August, but is most frequent in April and May. There is no delayed implantation and the gestation period is about six weeks. The young are weaned at four to five weeks, and there is normally a second litter. Young males of the first litter grow rapidly and are sexually mature by August, as are some of the females. Second litters grow more slowly and do not mature until their second year.

In Scotland the weasel has been known as the whittret, which is the equivalent of whitethroat in Suffolk. In Yorkshire it is known as the ressel; in Cheshire, the mouse-killer; in Sussex, the beale; and in some parts of Surrey as kine, which suggests Gilbert White's cane, the local name in Hampshire for 'a little reddish beast not much bigger than a field-mouse, but much longer', of his fifteenth letter to Pennant. The more general name weasel is from the Anglo-Saxon *wesle*, and was written *weesel* until at least the nineteenth century.

When, in 1892, Scotland suffered severely from a 'plague' of field voles, the Board of Agriculture appointed a committee of enquiry, and the examination of witnesses—farmers, keepers, shepherds—established beyond doubt that the weasel is the natural enemy of the short-tailed vole. The Committee voiced the opinion that it had been a grave mistake to destroy such large numbers. It was even suggested that weasels should be imported from the Continent and turned loose.

The voice is a guttural hiss when alarmed and a short screaming bark when disturbed, but neither is heard at all commonly.

The weasel is found throughout Britain, but is absent from most islands, and from Ireland.

Dental formula: $i \frac{3}{3}, \quad c \frac{1}{1}, \quad p \frac{3}{3}, \quad m \frac{1}{2} = 34.$

The Polecat

Family MUSTELIDAE   *Putorius putorius*

In contrast to the sweet-mart or pine marten, the polecat or fitchew was named the foumart or foul-marten, because the secretion from the glands under the tail is intolerably acrid and mephitic. On this account the fur is by some considered to be useless, but it was extensively used in the early nineteenth century, and to a slight extent still is today. Like the marten, the polecat, due mainly to the gin-trap and persecution by the gamekeeper, has become very rare in Britain, whereas formerly it was widespread. Its present stronghold is Wales, in parts of which it is now regarded as common. There have been reports of polecats from Devon, Cornwall, Gloucestershire, Herefordshire, Shropshire and the Lake District, but it is not always certain whether these have been feral polecat-ferrets. The destruction of the polecat may have been fully justified, in the interests of game preserves, but other interests would have been better served by its survival. With the decline in game preserves and a revival of re-afforestation a suitable habitat is provided by the Forestry Commission's plantations, and the increase of the polecat is encouraged as an insurance against vermin. It is still common throughout Europe, as far north as central Scandinavia.

Although in general appearance similar to the pine marten, the polecat is smaller, has shorter legs and a shorter tail, and differs in colour. A well-grown male is about 2 ft. (61 cm.) long, of which the bushy tail accounts for about 7 in. (180 mm.), and may weigh up to $4\frac{1}{2}$ lb. (2·05 kg.). The female has slightly less length and only a little over half this weight. In both sexes the long coarse fur is dark brown on the upper parts of the body, and black on the under surface. The head, also, is blackish, relieved with white marks about the muzzle and between the ears and eyes.

The usual habitat is a wood or copse, the polecat making its den in any suitable hole, such as a fox earth, rabbit burrow or natural rock crevice. A wood-stack has sometimes been used. With the approach of winter, polecats may seek shelter in deserted buildings. The polecat is less agile than the marten and less of a climber. It is active mainly at night.

There is a tendency, when speaking of polecats or any other carnivores, to stress that they are adept at finding entry into a hen-house, and have been known to kill all the occupants instead of taking just one. These, and other alleged misdeeds, tend to be emphasised and its normal diet, of rats, mice and rabbits, overlooked or ignored. Its diet also consists of eggs, birds, fish, frogs, lizards and snakes, including the viper, whose poison is considered to be innocuous in the blood of a polecat. Its usual method of carrying smaller prey is to grip it by the middle of the back, much as a retriever carries game. In addition to the remains of rats, mice, rabbits, birds and eels, there is one record of the discovery, in the larder of a polecat bitch, of the bodies of three kittens which were known to have been drowned at least $\frac{1}{4}$ mile away.

Doubtless accusations of poultry-killing are justified to some extent, but it is predators such as the polecat that keep the populations of small mammals within bounds. Without them, rats and mice, for example, would be a greater pest than they are, and the wholesale reduction in the numbers of polecats must have contributed materially to the calamitous increase in the rat populations. Since the polecats are mainly nocturnal, it should be possible to protect poultry from them. As to killing off all the hens in a roost, which usually evokes the castigation 'bloodthirsty', we have to remember that the hens cannot escape and nothing arouses the hunting instinct in an animal more than having its quarry panic. The mounting excitement in the victims stimulates an equally mounting excitement in the killer. It is the same with a fox and other similar animals.

Taken young, polecats are readily tameable, and one especial feature seen markedly in the tame polecats is their capacity for play. Young or half-grown polecats will tumble over each other in a bunch, endlessly wrestling and romping. In the course of this can be seen how supple their bodies are, when the fore-quarters are on the ground with the back of the neck touching earth while the hind-quarters remain in the normal position.

The polecat mates in early spring and the gestation period is about six weeks. From three to eight (mostly five or six) young to a litter are born in April or May, in a nest made of dry grass. It has been said that there is probably a second litter a few months later. At best this is uncertain, and it is unlikely since the young do not finally leave the parents until three months old, which would push the time of a second litter to late summer or early autumn.

Polecats are usually silent, but are said to use occasional short yelps, clucks and chatterings.

Dental formula: i $\frac{3}{3}$, c $\frac{1}{1}$, p $\frac{3}{3}$, m $\frac{1}{2}$ = 34.

The American Mink

Family MUSTELIDAE   *Mustela vison*

The rearing of mink on farms was established in Britain in 1929 and the industry has expanded greatly since World War II. From the beginning, mink were escaping, but there were no records of them breeding until 1957, when young mink were seen on the River Teign in Devon. Since then there have been widespread reports of breeding and there are now mink on many rivers in England, Wales and Scotland. Breeding colonies in the wild have been recorded in Devon, Hampshire, Wiltshire, Cardigan, Carmarthen, Pembroke and Aberdeenshire. Attempts are now being made to control their numbers by trapping.

Mink are reported to have killed poultry, pheasants, water fowl and domestic rabbits, as well as wild birds and mammals. Trout and crayfish are also taken. In Scandinavia feral mink are endangering the salmon stocks and in Iceland they have damaged wild birds, especially ground-nesters such as the waders.

The mink is similar in form to the stoat. The male is 17–26 in. (430–660 mm.) total length, of which 5–9 in. (130–230 mm.) are tail, the female being $\frac{1}{2}$–$\frac{3}{4}$ the size of the male. Weight is 1$\frac{1}{4}$–2$\frac{1}{4}$ lb. (0·5–1 kg.). Typically the fur is uniformly dark brown with a white throat patch, but colour varieties have been established by artificial selection. After World War II there were a number of reports of pine martens seen in southern England. Wherever these could be closely examined, as when they were shot and sent for identification, they proved to be feral mink.

The habits of these mink, which were imported from North America, are very similar to those of the European mink (*M. lutreola*). Nocturnal animals, they are solitary, each one having a den in a cleft of rocks or perhaps an enlarged water vole hole. Males range over long distances, but the females stick to a small area near home during the spring when they have four oestrus cycles between late February and early April. Ovulation only occurs after copulation. Implantation is delayed for a variable period of time and the litter of five to six are born, on average forty-five days (usually between thirty-nine and fifty-two) after mating.

Dental formula: $i \frac{3}{3}, \ c \frac{1}{1}, \ p \frac{3}{3}, \ m \frac{1}{2} = 34$.

The Feral Cat

Family FELIDAE   *Felis catus*

The records for the Penrhyn Estates, near Betws, in Wales, show that between the years 1874 and 1902, ninety-eight polecats, thirteen pine martens and 2,310 cats were shot on the game preserves. There is nothing to show how many of this surprising total of cats, an average of well over one per week, were domestic cats merely out hunting and how many were feral, i.e. cats truly gone wild. We can assume that many must have been feral, and breeding in the wild state, as it is difficult to see how this large total could be made up of pet cats in a district that could have been only thinly populated. None could have been the true wild cat for although this once ranged throughout Britain, by the middle of the nineteenth century it had been wiped out in England, Wales and southern Scotland. This left woods and forests, the natural home of the wild cat, ready for vacant possession by pussies taking it into their heads to live off the land.

The number of feral cats in the wooded parts of the countryside, especially on high ground, must be very high today. Occasionally one comes across such a cat, or even a nest of kittens among the bushes, well away from the nearest houses. For the most part, however, they lead their lives virtually unseen, resting up in trees, adept at taking cover at the first alarm, long before we can get near enough to see them. They are pariahs, with everyone's hand against them, and this must explain why practically nothing is known about them, except what can be gleaned from casual observation.

Usually it is the large domestic cat which takes itself off in this way, but some unwanted cats are taken out by car and abandoned at some lonely spot. Such cats, once they have gone wild, especially if they do so when young, grow larger than usual, probably as the result of a more athletic life as well as the plentiful natural food available to them. The only one I was able to study at close quarters, in a remote wood in Devon, seemed enormous, and its size was matched by its ferocity.

Captain Mayne Reid, the naturalist, wrote in 1889 of a tame cat in the Wye Valley that went to live in the woods and in four years doubled in size. This seems to be supported by such records as we have of the length of feral cats that have been shot. 3 ft. (90 cm.) long is common, and the longest measured 42 in. (1·06 m.) from nose tip to tail tip. This is nearly as large as the largest true wild cat, the maximum length of which is 45 in. (1·14 m.); and the greatest weight, 30 lb. (13·6 kg.), far exceeds that of the average wild cat which is 15¼ lb. (6·9 kg.).

The athletic abilities and ferocity of the wild cat are almost proverbial, but this is based on anecdotes rather than precise descriptions. There was, for example, the ancient and traditional story of the fight between Percival Cresare and a wild cat. The youth was returning from Doncaster through Melton Wood when, attacked by a wild cat, he sought refuge in the porch of Barnburgh Church, where the fight continued. In the morning youth and cat were found, the youth dead of severe lacerations, the cat crushed against the wall by his feet.

There is the better documented story of a female wild cat which leapt at a man's face when her kittens were nearby and only the vigorous use of a stick enabled the man to save himself from serious injury. These instances are exceptional because as a rule the cat, wild or feral, is more interested in vanishing into the undergrowth, and also because, in country where wild cats still live, people learn to give them a wide berth or go prepared to deal with any attack. They do, however, give us some insight into the character of the cat.

It is certain that domestic cats can more readily return to the wild than dogs, and when they do a natural ferocity, hitherto dormant, readily emerges. Take the remarkable case recorded in the *Field* for 1871, of a man who kept a number of tame birds and also a tame cat. For a long time the cat lived amicably with the birds. Then it started to kill them, one by one, and with this its natural instincts seem to have reasserted themselves for finally it leapt at its owner, landing on his chest, knocking him over backwards and scratching and biting him as he lay on the ground. His thick overcoat took the brunt of the attack, giving him time to beat off the cat, after which the animal departed and was not seen again. Possibly it went feral.

You can look around the English countryside and see no sign of feral cats, yet they are there somewhere, and as soon as you start enquiring the evidence is forthcoming, but only in bits and pieces. Gamekeepers are the best source of information, but they, as a class, know how to keep a still tongue. Even when they do talk you find their main interest has been to bury the carcase as speedily as possible. They do not take measurements of their kills, so for details of size you have to be content with 'enormous', 'big as a dog', and 'twice as big as an ordinary cat'.

A better account was given me by a friend recently. Lying on a bank watching for rabbits, he saw in the field beyond, a rabbit sail into the air and fall to the ground. Then he saw a large black fox, he thought, that threw the rabbit up again. But there was white on the animal's head so he thought it must be a badger. His field-glasses showed it was a huge cat, which he had under observation for forty-five minutes before, having eaten the rabbit and left only three tufts of fur on the ground, it went off into the nearby wood. My friend's assessment of size was that this cat made his own neutered tom 'look like a kitten'— and that tom weighs 21 lb. (9·5 kg.).

There seems little doubt that feral cats can equal true wild cats in size, possibly also in athletic performances. One has been described to me as leaping a $3\frac{1}{2}$ ft. (1·06 m.) fence clean, even carrying a rabbit in its mouth. This fits in with the stories of cats, tame, feral or wild, leaping at a person's face. To leap a fence with a rabbit is no less credible than authentic accounts of lions clearing fences, carrying the carcase of a buffalo, and other similar feats by the big cats.

It is said that these large sizes, and the intense ferocity, is reached by the second generation of 'gone-wild' cats. It is also asserted that all

(*right*) Adder (Note black zig-zag line on back)

PLATE 33

(*below*) Smooth snake

PLATE 34   (*above*) Chinese water deer   (*below*) Sika deer

revert to the tabby by the fifth generation, whatever the colours of their fore-bears. To be able to make such assertions with confidence means that somebody has a lot of information about feral cats that has never been recorded.

Dental formula: i $\frac{3}{3}$, c $\frac{1}{1}$, p $\frac{3}{2}$, m $\frac{1}{1}$ = 30.

The Wild Cat

Family FELIDAE   *Felis silvestris grampia*

The wild cat exists today mainly in the rocky parts of northern Scotland, north of the Great Glen. It is said to have considerably increased in numbers there in recent years. It inhabits the most lonely and inaccessible mountain-sides, hiding during the day among rocks, prowling far and wide at night in search of prey. It is of a general yellowish-grey colour, but there is a good deal of variation. Individuals differ in their dark brown markings, some having vertical stripes running down the sides, while in others these are broken up to form spots. Since there seems to have been considerable cross-breeding with feral domestic cats some of the variation in the pattern of the coat may be due to this. The wild cat has a squarish, robust head and a stouter and longer body than the normal domestic cat. The thick bushy tail is relatively shorter than in the domestic cat and is ringed, ending in a long black tip. The limbs, too, are longer than those of the tame cat, so that the wild cat stands higher. The fur is long, soft and thick.

The average length is about 2 ft. 9 in. (84 cm.), of which the tail accounts for 11 in. (28 cm.); but there is a record of a Scottish example measuring 3 ft. 9 in. (1·14 m.) in all. The weight of a male averages 11 lb. (5 kg.), but may be as much as 15 lb. (6·8 kg.), and that of the female 8½ lb. (3·8 kg.), with a maximum of 10 lb. (4·5 kg.), but one wild cat from the Carpathians weighed nearly 33 lb. (15 kg.).

Apart from the greater size and other differences given here in external appearance the wild cat is said to differ in features of its anatomy: in having a high arch to the nasal bones, larger carnassial teeth, and a gut only three-quarters the length of that of the domestic cat. The claws, also, are said to be horn-coloured.

12

Pennant (1776) says: 'This animal may be called the British tiger; it is the fiercest and most destructive beast we have; making dreadful havoc among our poultry, lambs and kids'. It is known to take mountain hares, grouse and rabbits, any small mammal or bird it can catch, as well as fish and insects. A wild cat can kill a roe fawn or well-grown lamb. The cat hunts alone or in pairs, mainly at night, with peaks of activity at dawn and dusk, except in autumn, when there is a tendency to hunt by day. Like the domestic cat, the wild cat shelters from rain but basks in the sun, on a bough or a rocky ledge. C. St John, writing in 1845, says that its strength and ferocity when hard pressed are perfectly astonishing. He adds: 'I have heard their wild and unearthly cry echo far into the quiet night as they answer and call to each other. I do not know a more harsh and unpleasant cry than that of the wild cat'.

The home range covers an area of 150–175 acres (60–70 hectares), which is defended by the male, who may wander well outside this in search of food in winter or in the breeding season.

The female makes a nest in some remote rock-cleft or hollow tree, away from the male who may kill the kittens. There are two breeding seasons in the year, in early March and again in late May or early June, the litters of four to five kittens being born in May and August respectively. Occasionally there is a third litter in December or January, but it is suspected that these may have been from a domestic-wild cat cross. Gestation is sixty-three days. The kittens have a light ground colour with greyish-brown tabby markings. They leave the nest at four to five weeks, but do not go hunting with the mother until ten to twelve weeks old and are not weaned until the age of four months. They leave the mother when five months old.

The voice is meeow, a growl when angry and a purr when pleased, with the typical small cat scream or caterwaul on occasion.

The distribution of the wild cat includes Europe and northern Asia to the north Himalayas. Though formerly widespread over the whole of Britain, and at one time a beast of chase in England, it appears never to have been a native of Ireland.

Dental formula: i $\frac{3}{3}$, c $\frac{1}{1}$, p $\frac{3}{2}$, m $\frac{1}{1}$ = 30.

Seals

Order PINNIPEDIA

Seals are carnivorous, related to such well-known carnivores as cats and dogs but sufficiently different from them to be placed in a separate order, the Pinnipedia (fin-footed). They are aquatic, their limbs are modified to paddle-like flippers, the tail is only a vestige, and the external ears are almost completely lost. The two species, the common seal and the grey or Atlantic seal, found around the British Isles, both belong to the group known as true or earless seals. They differ from the eared seals (sea-lions and fur seals) in being unable to bring the hind-flippers forwards. So although eared seals on land can use fore- and hind-flippers in locomotion, true seals can only progress awkwardly by a looping movement of the body, which is almost unaided by the flippers, the main thrust being imparted by the hips with the chest acting as a fulcrum. The external ear of eared seals is very small but can be seen. That of the true or earless seals is even smaller, little more than a flap of skin lying just inside the ear-opening and visible only when the fur is wet.

At one time seals of all kinds were killed for their fur, to satisfy the demands of fashion, and because the manner of killing was not publicised there was no stirring of the public conscience. Then came the period, not much more than a century ago, when seaside holidays became popular and more people became acquainted with live seals. This and other events have produced widespread changes in the attitude of many people towards seals and their exploitation and persecution. The growth of zoos and the increase in their popularity have brought us literally face to face with seals. Photographs and films of these animals are more widespread, so that we have become familiar with the face of a seal, with its large limpid eyes, with tears streaming down its cheeks.

The tears are the means whereby a seal rids its body of excess salt consequent on living and feeding in the sea. The eyes are adapted for underwater vision, and in spite of their apparently mute appeal are almost certainly myopic when the seal is on land. Even there, however, they serve some purpose. For example, a cow seal returning to her pup recognises it from a short distance by sight, mainly it would seem by

relating it to surrounding objects, since the cow confirms the identification by smell.

While a seal is submerged its nostrils and ears are closed, and in
murky waters, or at night or at depths of more than a few fathoms, we
may presume sight is no longer effective. Touch, presumably, is the
main sense, as in moles, possibly through the vibrissae (whiskers and
other bristles). No particular study has been made to test this, but two
observations suggest how these may operate. The first is that when a
seal has its head out of water, as for example when a grey seal is bottling
(floating vertically with head pointing upwards), the whiskers are laid
back along the face. As soon as the face touches water the whiskers are
brought forward, even more so when the seal starts to swim. This can
be compared with the action of a dog's whiskers. They are laid back
when the dog is still and brought forwards when it starts to advance.
In a dog, we may presume, the whiskers guard the face by making the
animal aware of obstructions, such as brambles.

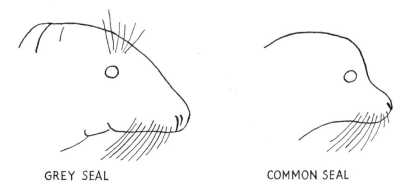

GREY SEAL   COMMON SEAL

*Profiles of grey and common seals, showing differences in shape of
head*

A seal may well use its whiskers to pick up vibrations in water, and it
may be they are most sensitive when erected. A seal at the surface, with
the lower part of the face awash, will have the whiskers erected. It will
take no notice of you or of a stone thrown through the air, but the
moment the stone touches the water, even a few yards from it, the seal
will dive immediately, 'like lightning'.

The male seals are called bulls, the females are cows, the babies are pups and their breeding grounds are rookeries. To add confusion to this already mixed nomenclature some authors have started the fashion of calling the babies cubs. Some pups are born with fur that is soon moulted, after which they are called moulters.

Common Seal

Family PHOCIDAE   *Phoca vitulina*

The common seal is known in North America as the harbor seal, a name more descriptive of its habitat. Except in the Shetlands and Northern Ireland, it hauls out on sand- and mud-banks, especially near estuaries. The bull is 6½ ft. (1·98 m.), the cow 1 ft. less (1·68 m.), their maximum weights being respectively 5 cwt. and 3 cwt. (253 and 150 kg.). There is little difference in appearance between the sexes, except that the bull has a broader and blunter muzzle. The coat is dark grey on the back, a lighter grey on the underside, and is mottled with numerous small black spots. The head is rounded, the nose 'retroussé' and the nostrils are set at an angle to the vertical.

More aquatic than the grey seal, it keeps mainly to shallow waters, and is sometimes seen proceeding by 'porpoise jumps' clear of the water. The usual times for remaining submerged are seven to ten minutes, five to six minutes for the pups, but adult common seals can stay underwater for up to forty-five minutes. When a seal submerges the supply of blood is cut off from all but the brain (and presumably the sense-organs) and the locomotory muscles, so conserving the supply of oxygen, making long periods of submersion possible.

Mating, which takes place in shallow water or at the water's edge, is preceded by rolling, blowing streams of bubbles, and slapping the surface of the water with the front flippers. The common seal is probably monogamous, possibly promiscuous, and there is no attempt by the bulls to establish a territory on the beach, as with most other seals. The breeding season varies slightly from one part of the British Isles to another, but is generally during late summer. There is a delayed implantation, until some time between November and January, and the pups are born in June, mainly. Birth takes place in the sea or on sand-banks or rocks exposed at ebb tide, and the pup is able to swim efficiently from birth. It needs some assistance from the mother, however, because for the first few days it is unable to haul itself on to the beach easily owing to the weakness of its flippers. A pup is $2\frac{1}{2}$–3 ft. (70–90 cm.) long at birth and weighs 20–25 lb. (9–11 kg.). A few are born with a white coat, which is lost within two days, but usually the coat is grey at birth, similar to that of the parents. The baby seal is suckled underwater and is weaned at three weeks to a month.

The common seal feeds on fish and shellfish. Species known to be taken include dab, plaice, flounder, whiting and goby, as well as various molluscs, shrimps, prawns and crabs. Feeding times vary from place to place. In the Wash feeding is during the day-time tides, the seals hauling out at night. In the Shetlands the common seals haul out at low water and feed at high water both by day and night.

The voice is little used, but the adults growl when cornered or yelp and howl like dogs when excited. The pups mew.

Common seals range throughout the North Atlantic and North Pacific, as far south as France in the east and the New England States in the west, and south to California and northern China in the Pacific. In the British Isles breeding colonies are found on the coasts of East Anglia and the Wash, eastern Scotland, the Orkneys and Shetlands, the west coast of Scotland, Inner and Outer Hebrides and Northern Ireland. Outside the breeding season there is considerable movement, so that common seals may turn up anywhere around the British Isles. It is possible also that there is an interchange between the seals on our east coats and those on the Dutch and Danish coasts, but this has yet to be verified. The numbers of common seals are difficult to assess, for the same reason that there are so many gaps in our knowledge of this species, that so much of its time is spent in water, making it inaccessible for study. Some idea of numbers can, however, be gained from the estimates of about 2,000 in the East Anglia–Wash area, 350 in the Orkneys and 400 in the Shetlands.

The dental formula is: i $\frac{3}{2}$, c $\frac{1}{1}$, post-canines (the equivalent of pre-molars and molars) $\frac{5}{5}$.

(Only the second, third and fourth post-canines have precursors in the milk dentition.)

Grey or Atlantic Seal

Family PHOCIDAE   *Halichoerus grypus*

The grey seal, sometimes called the Atlantic seal, is not confined to the western shores of the British Isles. One of the best-known colonies is the Farne Islands, off the coast of Northumberland and a few grey seals have moved as far south as the Wash, on the east coast. It spends much of the year at sea, coming ashore to breed in autumn. It also hauls out on rocks and small islands at other times, when moulting or when feeding inshore. When ashore for breeding the colonies contain adult males and females as well as sub-adults, but at moulting time there is a segregation of the sexes because male and female moult at different times: the cows from January to March, the bulls from March to May.

The bulls are black or dark brown to grey, with lighter spots and patches, up to 8 ft. (254 cm.) or more long, exceptionally up to 9 ft. The cows are smaller, up to 7 ft. 3 in. (220 cm.) and light grey to fawn, with a conspicuously lighter underside and with darker spots. But colour is unreliable as a means of identification since both bulls and cows appear darker when wet. The better means of identification are the flattened top of the head, the almost vertical nostril slits and the convex profile of the snout, pronounced in the bull, almost straight in the cow. Mature bulls also have much blubber on the neck and shoulders, so that the skin is thrown into folds, and they have a wider muzzle, but young bulls nearly resemble the cows in having the less heavy fore-quarters.

In swimming the body is propelled by side-to-side sweeps of the hind-flippers, the fore-flippers serving as balancers or for paddling when the seal is idling, usually inshore.

In July and early August the bulls begin to move towards the rookeries, followed by the pregnant cows. On land the bulls take up territories which they defend against later arrivals. The cows bear their pups, suckle them for two to three weeks then desert them, to mate, an old bull serving anything up to twenty cows, on average a dozen. After this the cows leave the rookery to start feeding, but the bulls maintain their territories for another six weeks to two months. Cows do not become sexually mature before four years of age and bulls not until six, and at the breeding season these immature individuals occupy separate beaches from the breeding bulls and cows.

The pups are born on beaches backed by cliffs or on rocky ledges, and in some places on grassy level cliff-tops as much as 50 ft. (15·2 m.) above sea-level, reached by rocky gullies. The bulls are to the seaward of the pupping grounds. The cows travel to the sea to feed and return to suckle their pups, travelling along the gullies to do so. In the Outer Hebrides, where there is risk of pounding from the Atlantic swell, the cows do not feed for the two weeks' suckling period. The pups are born with a creamy-white coat which is moulted at three to four weeks, and replaced by light grey to blue-black fur. They are up to 3½ ft. (107 cm.) long and weigh up to 42 lb. (19 kg.). They grow rapidly, increasing at the rate of 3–4 lb. (about 1·5 kg.) a day, the mother's milk being very rich, containing 60% of fat compared with the 3½% in cow's milk, which enables them to lay down a thick layer of blubber. At birth the pup's skin is loose and lies in folds, but this is soon filled out. By the third week, when moulting begins, the pup will have doubled its weight. The moult begins on the muzzle and hind-flippers. Some pups lose their white puppy coat in patches, others lose it more uniformly, according to whether they enter the water or not during the moult. In those that do the hair is washed off and the moult appears more uniform.

Whether pups enter the water or not during the moult is, like so much else in the behaviour of grey seals, a matter of geography. In some areas moulting may be patchy or uniform, in others there may be a predominance of one or the other. It is not possible to do more than generalise here because to do more would require particularising for almost each group of islands or beaches used by the seals.

Development of the pups seems to be rapid. Birth may take one to fifteen seconds; growth is rapid, so is the moult. The pups are soon able to swim and to fend for themselves, and in this they are helped by the speed with which their second teeth appear. The milk teeth are absorbed before birth, the permanent dentition erupting rapidly before or after birth and in any case before the pup is two weeks old.

At the age of four weeks, except on such islands as North Rona, where they go farther inland and remain ashore for a longer period, the pups take to the water, possibly hauling out from time to time. There is, however, a high mortality among the pups due, it has always been said, to starvation, injury resulting from fighting between bulls that charge across the beach regardless of whether there are pups in the way, and to storms. It now seems there is a revision of opinion on this, however, and that the deaths are due to starvation and disease. But details are lacking and further research is needed.

Although nearly a year elapses between mating and the birth of the pup, the actual gestation is no more than seven months. There is delayed implantation. The embryo undergoes the first stages of development then lies dormant in the uterus until after the cow has completed her moult.

There is little precise information on the food taken except that it includes crabs and lobsters, squid and cuttlefish, and a variety of fish, such as lumpsucker, conger, saithe, pollock, mackerel, herring, pilchard and salmon.

Only 20% of pups reach maturity, but once this is attained the survival rate increases steeply. Cows are longer-lived than bulls, with thirty-five years as the maximum recorded, the record for a bull being twenty years.

When at sea grey seals are silent. They then use a smack of the hind-flippers on the surface as they dive for an alarm note. On or near the shore they are vocal. The pups make a hissing snarl when approached and a high-pitched bawl, like a child, when hungry. The bulls hiss and snarl, and the cows have a characteristic singing hoot which, when a chorus starts up, sounds like moaning at a distance.

Grey seals are found in small numbers in the Baltic, on the coast of Norway, on Novaya Zemlya and Murmansk, the Faroes, Iceland, the southern tip of Greenland, Nova Scotia, Newfoundland and the Gulf of St Lawrence. Occasionally they go as far south as the New England States, in America, and they are found sparingly on the north coast of France. They are also listed as one of the marine mammals of Portugal.

Only in the British Isles are they protected. Elsewhere they are hunted for their skins, oil or flesh, or harried because they are supposed to be detrimental to fisheries or for the damage they do to salmon nets. Out of an estimated world population of 45,000 all but 10,000 live around the British Isles.

The dental formula is: i $\frac{3}{2}$, c $\frac{1}{1}$, post-canines (equivalent to pre-molars and molars) $\frac{5}{5}$.

(Only the second, third and fourth post-canines have precursors in the milk dentition.)

CLOVEN-HOOFED ANIMALS

The Red Deer

Family CERVIDAE   *Cervus elaphus*

Indigenous red deer are now mainly confined to the Scottish High-lands and Islands, the Lake District and Devon (Exmoor, Quantocks and Brendon Hills), but feral red deer, from introduction or escapes from park herds, are present from the south-west counties eastwards to Sussex (Ashdown Forest), and also in Thetford Chase (Norfolk), the Roches (North Staffordshire), Cheshire, Lancashire, West Yorkshire and Durham, as well as the Scottish Lowlands and Co. Donegal and Co. Wicklow in Eire. Outside Britain red deer range across Europe and Asia, as far south as the southern slopes of the Himalayas.

The fully grown stag stands up to 4 ft. 6 in. (1·4 m.) at the shoulder. The hind is somewhat less. The summer coat is reddish-brown, some-times golden-red, and changes to a brownish-grey in winter by the new growth of grey hairs. On the underparts the colour is white, and a patch of white around the short tail furnishes a 'recognition mark', common to most of the deer family, which apparently serves in guiding other members of the herd when in flight from a predator. The usual gait is a steady trot, the gallop being used only in moments of alarm or near danger, which is when the tail is held erect exposing fully the white 'signal'.

The true home of red deer is in dense woodlands. They are browsers rather than grazers, and enforced changes in feeding habits have led to a wide variation in both body size and size of antlers under present-day conditions. In Europe, where the red deer inhabits forests and can enjoy the largest amount of browsing, the maximum weight goes up to 40 st. (254 kg.). In English woodlands it does not exceed 30 st. (190 kg.) and on the Scottish moors, where the feed is poor and is mainly grass, it is seldom more than 15 st. (95 kg.). Forest deer also have larger and heavier antlers than those subsisting on grazing.

The function of the antlers is still problematic, but the likelihood is that they are more an indication of social rank than any form of defence. We know that in the larger deer, such as the elk and moose, defence against predators is by the animal rising on its hind-legs and striking down sharply with the front hoofs. The killing of a hound by the antlers of the red deer at bay is more to be regarded as misadventure, the hound having leapt on to the antlers rather than the stag deliberately thrusting with them. The antlers are used in fights between rival stags, but even in such combat the main force of the encounter is in butting with the forehead. Sometimes one of the protagonists may be impaled on its rival's antlers or both sets may be interlocked so that the two opponents eventually starve. Such accidents are rare, however, and to be regarded as incidental to the fight.

The antlers are cast between February and April, the young ones casting theirs later than the older stags. The coat also is moulted in May. In spring and summer while their antlers are growing the stags live apart, solitary or in small groups, but in early autumn when new antlers are fully grown, clean and hard, the stag's 'belling' call to the hinds, or challenge to a rival, may be heard. From early September to mid-October the stags visit increasingly the peat bogs and muddy pools to wallow. There is a good deal of furious fighting—almost mock-fighting —when two stags of similar age and strength meet in the vicinity of the hinds. The stag is then in prime condition. His neck and shoulders become clothed in a thick mane of long brown hair just prior to the rut and his head becomes adorned with antlers that reveal his age. It has usually been assumed that these adornments serve to make the stag attractive to the hinds, that is, the finer the antlers the more their possessor succeeds in building up his harem. One has only to watch a herd in the rutting season to see that the hinds seem to be quite in-different to the stag's antlers, his rampaging rushes round the perimeter of the harem, and to his bellowings. They graze unconcernedly while he challenges or fights a rival, and a hind will accept a young stag entering the harem while the harem-master is engaged in trying to keep another rival out on the other side. Certain stags, known as hummels, are without antlers, yet they form harems, fight rivals and breed as successfully as the antlered deer. Some observers maintain that, if anything, they breed more successfully.

The points on the antlers increase in number with age and when twelve are present the stag is known as a royal. The first sign of antlers is the growth of two hair-covered pedicles, or knobs, as a result of which the young stag is then known as a knobber. During the second year of life a stag has a pair of simple un-branched stems, when it is known as a pricket. In the third year the new antlers bear a tine or branch and an additional tine is added each succeeding year, until the normal head of antlers is reached, after which growth becomes somewhat irregular. The first tine to appear, in the third year, is the brow tine. The next is the bay or bez tine. The third is known as the tray or trez tine. After this the antler achieves three branches at the top, which are known as the tops. At this stage a stag is said to have 'all his rights' (i.e. brow, bez, trez and three tops). On the main beam between the bez and the trez should be a knob, known as a snag, and between the trez and the tops is another, known

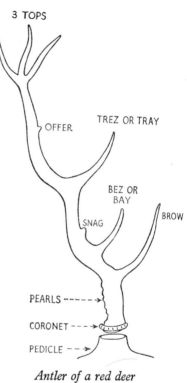

Antler of a red deer

as the offer. The royal is sometimes exceeded and antlers with more than twelve points can be very heavy, the record for the species being 74 lb. (26·2 kg.).

The hinds are sexually mature at three years of age and bear their first calves at four years. Gestation is eight months and the peak of the calving is from the end of May to mid June. Then the hinds separate, each retiring to a lonely spot among the bracken where her single calf, very occasionally two, is dropped. A calf is born already covered with fur, and its back and sides are dappled with white after the manner of the fallow deer, but the spots on the red deer are not retained beyond calfhood; they begin to fade at a month, and are completely gone at two months. It is able to stand within a few minutes of birth and can run within a few hours. On the other hand, it does not feed itself until eight to ten months old, and remains with the mother until the following autumn. The longevity of red deer is about twenty years.

Red deer are grazers and browsers, feeding on grass and the young shoots of trees and shrubs, but, as we have seen, those unable to browse suffer extensively from weight and size of antlers. In addition to the damage to the plantations of young trees, agricultural land can be harmed badly by them, a whole field of turnips being ruined in a night by a visit from a herd of deer. They will also destroy wheat, potatoes and cabbages. Additional natural foods include toadstools, acorns and chestnuts. Seaweed is eaten when the deer are living near the coast. Wild fruits are taken in autumn, but little is eaten by a stag during the rut. Feeding is mainly at dawn and dusk.

In Britain today the only enemy of adult deer is man, but the calves may be taken by foxes, wild cats and eagles.

Red deer are fairly silent. The hind barks a warning, or gives a nasal bleat when alarmed and the stag uses a gruff bark, but less frequently. The best-known call is the roar of the stag during the rut, a terrifying sound when heard at close quarters, and the hind may occasionally roar in summer. The calf will call, when distressed or alarmed, with a high-pitched bleat or scream.

The teeth of the red deer do not differ materially from those of the ox and the sheep. There are no teeth in the fore part of the upper jaw, the three pre-molars and three molars of each side being placed well back in the cheek. In front of each half of the lower jaw are three incisors which bite against the hardened gum in the upper jaw. The stag alone has a single canine tooth a little behind these. Three pre-molars and three molars correspond with, and bite against, those of the upper jaw. Dental formula: $i\ \frac{0}{3}, \quad c\ \frac{0}{1}, \quad p\ \frac{3}{3}, \quad m\ \frac{3}{3} = 32.$

The Roe Deer

Family CERVIDAE   *Capreolus capreolus*

The roe is found mostly in open woods. It is the smallest and most dainty of our native species, and appears to have been formerly the most widely distributed of the three, although absent from Wales and Ireland. It would appear to have been driven farther and farther north by the increasing human settlement in the south. As a truly wild animal it had disappeared from England, but there have been a number of reintroductions to forests in widely separated parts of the country of sixteen counties. There, although in places very numerous, its ability to move stealthily through dense cover makes the chance of seeing it remote, more particularly as the roe is largely nocturnal in its habits, feeding mainly at dusk and dawn. In moderately well-populated areas of southern

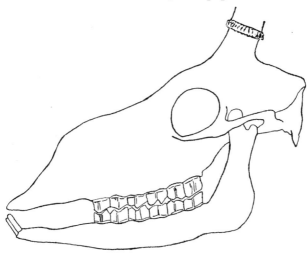

Skull of roe deer

England some inhabitants often see roe in their gardens in the early morning, but more commonly found are the numerous hoof-marks (slots) in soft ground.

A well-grown roe buck stands about $2\frac{1}{2}$ ft. (76 cm.) at the shoulder and weighs up to 70 lb. (32 kg.); the doe, being smaller, does not exceed

46 lb. (21 kg.). In summer the coat is bright red-brown, short and smooth; but in winter it becomes long and brittle, and the colour changes to a warm grey. The changes occur in April to May and October to November. The tail is so short as to be barely visible among the surrounding hairs which, as well as the underparts and the inner sides of the thighs, are white. The ears are relatively larger than those of the other species, covered with long hairs, and whitish inside. It has a white chin and a white spot on each side of the dark muzzle.

There are no signs of antlers in first year fawns; in the second year they make their appearance as simple unbranched prongs. The third year the antlers are forked, a short tine pointing forwards; those of the fourth year have an additional tine directed backwards, and this marks the full extent of their growth. There are no brow tines. In later years they have the same general design, although they grow larger each year, but even at their maximum they are only 8 or 9 in. (200–230 mm.) long, and are nearly upright. Small though these antlers are, they can be dangerous, if only accidentally, and they have on occasion caused fatal injuries to human beings. They are shed in November to December. The new antlers are fully grown and clear of velvet by April or early May.

Roe deer are monogamous. They never congregate in large herds, but form small family groups, which persist until the end of the winter when the young are driven away. In spring the doe retires deep into the covert, where her two, rarely one or three, fawns are born. They are spotted for the first year only. When they are about a fortnight old she brings them out into the more open parts to join the buck.

The rut takes place in late July or August, the bucks barking loudly and the does uttering a squeaking cry. Implantation is delayed and takes place in late December. The fawns are dropped in May or June of the following year, after a gestation of five months. A feature of the rutting season, which has attracted a good deal of attention in recent years, is the use of 'roe rings', in which a form of courtship takes place. The place chosen is a piece of open ground around a tree, bush or clump of vegetation. This is marked by the buck fraying the young trees, barking them, and scraping the ground, possibly marking the scrapes with scent from glands on the forehead. Around this central point the

PLATE 35   (*above*) Sand lizard sloughing

(*centre*) Grass snake
(Note yellow 'collar' on neck behind head)

(*below*) Two adders, one with normal colours, and a melanistic
(black) individual on right

PLATE 36 (*above*) Smooth newt (male) (*below*) Crested newt (male)

buck chases the doe, their hoofs wearing a ring or figure-of-eight in the
ground. The doe appears to be a willing partner although the impression
is that the buck, who follows her closely, drives her round. He often
appears, also, to be heading her back into the ring. In some places they
have been seen to run round and round the ring; in other places they
merely walk. Mating takes place in the ring. A pair of roe remain
together, at least for the season.

Roe buck chasing doe in a 'ring'

The roe is a good swimmer, and often crosses rivers, probably in
order to get a change of food, though sometimes there is no apparent
reason. They have been known to swim across wide pieces of water, and
even arms of the sea.

Leaves are browsed and berries eaten, as well as fungi and clover.
Bark is rarely eaten. Foods range from the foliage of broad-leaved trees
and shrubs to yew and pine shoots, heather and juniper, as well as briar,
bramble and privet.

Outside Britain, the roe extends across Europe, from the Mediterranean in the south to the level of southern Scandinavia in the north,
and thence into Asia.

Dental formula: i $\frac{0}{3}$, c $\frac{0}{1}$, p $\frac{3}{3}$, m $\frac{3}{3}=32$.

13

The Fallow Deer

Family CERVIDAE   *Dama dama*

The fallow deer is readily recognisable by its palmate antlers, flattened and expanded in all the branches of the upper part, though the main stem or 'beam' is rounded as in the red deer. The buck only wears antlers, and these have the brow tine and the trez, but no bez. The main part of the antlers forms a broad curved plate, the margins of which are drawn out into a number of flat points. Its description 'palmate antler' refers to the resemblance to the palm of the hand with its finger prolongations. The antlers are shed annually in May, and the new antlers are fully grown and clear of velvet by the end of August. There are no upper canine teeth and the first upper incisors on each side is expanded.

The fallow deer is smaller than the red deer, the buck standing at most 3 ft. (91 cm.) at the shoulder, and the hind somewhat less. It also differs in colour from the red deer, being a paler red or reddish-yellow above, spotted with white, with yellowish-white underparts. The tail is longer than that of the red deer, and is kept in constant motion from side to side. The vertical white stripe on either side of the rump shows up strongly when the animal is in retreat. This is usually regarded as a signal or mark enabling those following to keep in sight the one leading the headlong dash. It is more likely an alarm signal, the hairs, being erected, making the stripes show more white, thus flashing the alarm from one to the other through the herd.

In winter the fur is a greyish-brown, the change taking place in October. Some of the park herds represent a darker race and show this dark colouration at all seasons. It has been suggested that they are descended from a darker, hardier race introduced from Norway by James I; but one authority claims this variety was in Windsor Park as far back as the year 1465. The dark form may be seen in Epping Forest, also in Richmond Park and Bushey Park, where, however, the lighter form is in the majority. The fallow is a native of southern Europe, its range extending along the Mediterranean border and into Asia Minor. It is said to have been introduced into Britain by the Romans, though fossil remains found here suggest that it may have been a true native

originally. Nevertheless, not all authorities agree and it is still an open
question whether these fossil remains represent the fallow deer.

During the first year the fallow fawn gives no sign of antlers, but in
its second it produces a pair of short unbranched prongs, which give
the fawn the name of pricket. The next year the simple prongs are
succeeded by antlers that bear two forward tines, and the extremity of
the beam is slightly expanded and flattened, and its margin indented.
In the fourth year the form of the antlers is similar, but they are more
developed, the flat portion of the beam being much larger and its outer
margin more regularly toothed or snagged. The fifth year shows further
advance along the same lines, and the animal then becomes known as a
'buck of the first head'. In later years the additions are represented merely
by an increase in the number of spillers or snags to the flattened beam.

In October the bucks gather the does into harems, and in the course
of this there is a certain amount of fighting, but this is mainly show.
The old bucks mark the rutting area to be defended by fraying the young
trees with their antlers and urinating in scrapes beneath them. The
young bucks establish territories downwind of these. Throughout the
winter fallow deer may be encountered in mixed herds of both sexes. At
other times the sexes are segregated, the mixed herds reforming after
midsummer. In winter fallow deer not only browse young shoots but
the does more especially kill many small trees by eating their bark. They
will also eat acorns, chestnuts and horse chestnuts. Their main food is,
however, grasses and herbs. Feeding is mainly at dawn and dusk although
fallow deer can be seen about in the day-time, particularly in winter,
apart from the old bucks which tend to be mainly nocturnal and rarely
seen.

The rut lasts about a month, during which does of two years or older
become pregnant. The fawns are born in May or June far in among the
bracken, following a gestation of eight months. There is mostly only
one at a birth, occasionally two, but rarely three. The fawn is able to
run within a few hours of birth, but its tendency is to use this only in
emergency. When approached the fawn lies perfectly still until the last
moment, when it bounds rapidly away.

Although only the bucks have antlers, it is the does that protect the
herd. It is they that alert it and when it moves off we see, as likely

as not, when the herd is caught in the open, the massed antlers of the bucks huddled at the centre, surrounded by a screen of does. As the van of the herd draws near to cover, the ranks of the does open and the bucks, with one accord, take the lead in seeking the security of cover. The fallow's preference is, however, to seek security in thick cover rather than in flight.

The name fallow is the Anglo-Saxon *fealewe*, and indicates the fulvous colour of the lighter race.

Dental formula: i $\frac{0}{3}$, c $\frac{0}{1}$, p $\frac{3}{3}$, m $\frac{3}{3}$ = 32.

The Sika

Family CERVIDAE   *Cervus nippon*

The sika is somewhat smaller than the fallow deer with a similar coat. In summer the flanks are a warm buff-brown with faint spots, changing to a uniform darker brown in winter. The pure white tail enables it to be easily distinguished from the fallow deer. When flushed, the rump is seen as a white heart-shaped patch with black borders. The antlers are like those of the red deer but simpler, with never more than four points. They are cast in early April, are in velvet from May to July and are cleared by September.

In habits the sika is similar to the red deer but is less gregarious, usually travelling singly, or in small groups especially towards the end of winter when up to twenty stags may be seen together. Hinds and calves also gather in groups throughout the winter. They feed in open woodland, heath, moor or, preferably, grass at night, having spent the day in thick cover. The usual food is rough herbage and grass, but sika will also strip bark, especially of hazel.

During the rut the stag has a characteristic rising and falling whistle that ends in a grunt. This is repeated three or four times, followed by a long silence of up to an hour. When alarmed, a sharp scream ending in a grunt is given. The calf is born in late May and is very similar to that of a fallow deer.

Sikas were imported from Japan in the mid-nineteenth century.

Escapes from parks, mainly during the two World Wars, have resulted in feral herds becoming widespread in southern and northern England, in Scotland and in Ireland.

Dental formula: i $\frac{0}{3}$, c $\frac{0}{1}$, p $\frac{3}{3}$, m $\frac{3}{3}$ = 32.

The Chinese Muntjac

Family CERVIDAE   *Muntiacus reevesi*

Two species of small deer, known as muntjac, have been liberated in England, but only the Chinese species is now at large. Indian muntjacs were turned out at Woburn Park some seventy years ago. Later the Chinese species was also brought over to Woburn and it is this that has become well established as a wild animal, although there has been some interbreeding with the Indian species. From Bedfordshire, they have spread into the neighbouring counties and have been found eastwards to Norfolk and Suffolk and southwards to Middlesex and Dorset. The exact extent of the muntjac's spread is not fully known. Nowhere is the population dense and, although diurnal, they are not often seen as they are well concealed in dense undergrowth and bracken. Their numbers also cannot be fully assessed for the same reason and also because of fluctuations. For example, severe winters cause a heavy mortality through starvation. However, feeding takes place in more open country in the morning and evening. The presence of muntjacs in an area can be detected by observation of the frayed trees along their regular runs that lead to feeding places and latrines, and by their odd tracks. The two halves of each foot are uneven in size.

The call is a short, dog-like bark (hence the alternative name of barking deer) or a rhythmical clacking noise given when running. This is believed to be a succession of barks made by rhythmic expulsion of breath.

Muntjacs are small, reddish deer, about 16 in. (40 cm.) at the shoulder and, characteristically, they stand with head lowered and back rounded. The short antlers, which are in line with the slope of the forehead and are mounted on long pedicles, have only two points. The fairly long tail has a bushy tip. The stags have long tusk-like upper canines. The face is marked longitudinally by two slit-like openings of facial scent glands, hence a further name of rib-faced deer.

The bucks and does associate from October to March; otherwise the muntjacs are largely solitary. One, sometimes two, fawns are born, mainly in late summer, but there is evidence of breeding all the year round.

Dental formula: $i \frac{0}{3}$, $c \frac{1}{1}$, $p \frac{3}{3}$, $m \frac{3}{3} = 34$.

Chinese Water Deer

Family CERVIDAE $\qquad$ *Hydropotes inermis*

This is the smallest species of deer at large in Britain and was introduced at Woburn at the turn of this century. It is now widespread in Bedfordshire, Buckinghamshire and Hertfordshire. Stock from Woburn introduced to Hampshire and Shropshire is now also going feral.

The coat is pale brown stippled with black. Although slightly smaller than the muntjac, they are readily distinguished by the absence of antlers and the posture of the head. Both sexes have long, movable upper canines, though those of the female are rather shorter than the male's.

Little is known about the habits of the feral stock. They are solitary and diurnal, and observers speak of them having a bounding run 'like a rabbit'. In their native China they inhabit swamp country, but in England they live in dry woodland and open country. The fawns are born in May or June. The size of the litters is unique amongst deer. In China they consist of five or six fawns, though there are less than this in English parks.

Dental formula: $i \frac{0}{3}$, $c \frac{1}{1}$, $p \frac{3}{3}$, $m \frac{3}{3} = 34$.

The Goat

Family BOVIDAE $\qquad$ *Capra hircus*

Goats have no doubt been escaping and going wild ever since they were domesticated, but apart from these, some have deliberately been put out. At one time it was the custom for Welsh herdsmen to let light-coloured goats run with the dark cattle so that the whereabouts of the herds could be more readily seen from a distance. Further reasons for the inclusion of goats with domestic stock were that they were said to prevent contagious abortion of cattle, to attack and kill adders and because goats were supposed to be more sensitive to the onset of bad weather. Other goats were liberated in the Welsh mountains to browse the rich grass growing on inaccessible ledges. Sheep were able to get up to this vegetation, but often could not get down and had to be retrieved laboriously, whereas the goats were more likely to be able to make the return journey. By eating the lush grass the goats removed from the sheep the temptation to stray.

Such goats revert to the wild type within a few generations. They live in small herds from which the mature males keep away except during the rut in autumn. The kids are born in spring.

Feral goats are to be found in central and north Wales, Cheviot, Northumberland, the Scottish Highlands and on various islands such as Lundy, Holy Island, Rathlin Island and Achill Island in Ireland, and in some of the Hebrides.

Dental formula: $i\frac{0}{3}$, $c\frac{0}{0}$, $p\frac{3}{3}$, $m\frac{3}{3}=30$.

LIZARDS

The Common Lizard

Family LACERTIDAE   *Lacerta vivipara*

It is a pleasant occupation to sit on a sunny, heather-clad hillside and, keeping perfectly still, watch the antics of the common lizard. We first become aware of it peeping at us from under cover. If it is not actually peeping at us, it is at least watchful for any unusual movement. Then, if we remain quiet, we may see it leaping swiftly over the crowded plants. Its movements are so rapid that it is not at all easy to follow them in detail. Indeed, the movement of the lizard resembles more closely a mere shadowy streak. It is even more difficult to catch one for closer examination. The lizard runs with a nimble, gliding motion; the body and tail are scarcely lifted from the ground. The usual mode of progression is to shoot forward horizontally from one tuft of herbage to the next, but they also run with facility over the shoots of heather or heath, their long, delicate toes ensuring as safe a landing as that of the squirrel leaping from branch to branch. But whereas the squirrel clings with its claws, the toes of the lizard are merely spread out to cover the gaps between the foliage. They are not used for gripping. Nevertheless, the claws can be used to ascend vertical walls or even posts with smooth surfaces. The common lizard can also swim well and will readily pursue prey in water.

A common habit is to seek a patch of sand fully exposed to the sun and there bask. The lizards seem to have favourite spots where they can be found day after day. Yet in spite of this habit of basking, lying with the body flattened and limbs extended as if to catch as much of the sun as possible, the common lizard is intolerant of excessive heat. Individuals in captivity soon die if exposed to too much sunlight.

The common lizard averages about 5 in. (128 mm.) in length and the maximum attained by a male is just under 7 in. (179 mm.) and by a female slightly over 7 in. The females are therefore little, if any longer, and doubtless, a longer series of measurements would show even less difference between the sexes. The female is, however, the more heavily built; the male is the more graceful of the two, his tail tapering gradually from the slender body to the very fine tip. Though the tail is in both sexes equal in length to the head and body, that of the female appears shorter, owing to its sudden tapering beyond the thick basal portion.

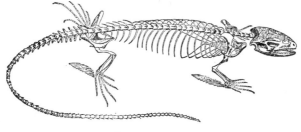

Skeleton of lizard

The colour is some tint of brown, varying considerably in different individuals from yellow-grey to purple-brown, as a ground tint, upon which are laid variable dark spots forming more or less broken longitudinal lines. There is sometimes a blackish line or band following the course of the backbone to a little behind the hips, and a dark band along the sides edged with yellow. On the underside the males are orange, or red, spotted with black; the females are orange, yellow, or pale greenish, with or without black spots, sometimes with only a few small grey dots. They appear to moult or 'slough' in patches, though entire sloughs are found occasionally.

Their food is chiefly insects, including flies, some beetles, moths, as well as ants and their grubs, but spiders are probably even more readily eaten. Small caterpillars are swallowed whole, larger prey may be shaken as a terrier shakes a rat. Large caterpillars are chewed, their insides swallowed and the skin rejected.

Mating takes place in April and May, with no obvious courtship but with some fighting between the males. Gestation is about three months.

The name *vivipara* refers to the fact that the female retains her eggs until they are fully developed and ready to hatch, so that the young are born free from the egg-membrane, or else the membrane is broken in the act of oviposition or immediately after. It is often said that they are deposited anywhere, with no attempt at a nest or concealment, and that the mother exhibits no interest or concern in her progeny. This appears to be true of those kept in captivity, but in the wild the female digs a shallow pit, preferably well concealed in moist soil, into which she

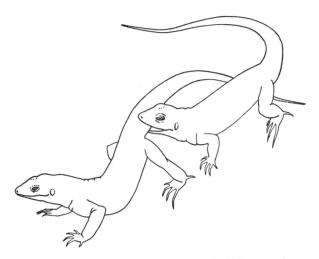

A feature of the lizard courtship: male biting female

deposits her young, in July or August. These number from five to eight in a litter, exceptionally four or ten, and are 1½–2 in. (38–50 mm.) long. Most of them are bronze-brown but a few are black. These change to bronze-brown within a week. The underparts are greyish-brown, and often the back and sides are speckled with gold. Within a few hours they begin to feed, hunting small insects such as aphids and other soft-bodied species. From the first they are agile and skilful in the search for food, and do not attempt to take prey larger than they can manage. The teeth are very small and conical, and unfitted to deal with hard substances. As the two halves of the lower jaw are firmly connected there can be no distension of the small mouth to accommodate large prey, as happens with snakes. Sexual maturity, in males at least, is reached at twenty-two months.

The usual attitude of the common lizard is with the extended tail and greater part of the body resting on the ground, or other support, whilst the head and fore-parts are raised on the arms, and the muzzle turned to one side in an attitude suggestive of listening. Lizards are said to be susceptible to musical sounds, and there are those that claim they can attract lizards from their hiding places by a particular whistle.

One of the best-known characteristics of the lizard is the brittleness of the tail. When catching, or attempting to catch a lizard, it is best to grasp it by the shoulders. If the tail is held instead, it will probably come away in the hand, snapping readily at a joint near the base, as though it were glass or sealing wax. A tail will grow from the stump if the lizard lives long enough, but it is always a poor, ungraceful affair compared with the original. The general idea is that the tail snaps off as a result of the mechanical pressure exerted in grasping it. In fact, the lizard throws off its tail. For example, it can happen that you grab the lizard by the body, yet the tail comes away, nevertheless. Conversely, lizards accustomed to being handled, and tame enough to have lost any sense of alarm at the close presence of human beings, will not lose tails even when held by them. The autotomy, as the operation is called, is governed by a nervous reflex. The tail itself is provided with a special breaking plane. At that point, there is a line of weakness through one of the vertebrae and opposite this the blood-vessels and nerves are narrowed as in the waist of an hour-glass. As a consequence, everything is ready for the fracture with the least damage and shock to the animal itself, when the tail is thrown off by muscular contraction. The narrow neck in each blood-vessel at this point results in a natural ligature, when the break occurs, thus obviating undue loss of blood.

The tail when cast off continues to lash violently, so that it twists and bounces in a truly startling manner. The eye is held by this unusual spectacle, while the former owner glides like lightning into the nearest cover. It is easy to see that a predator on the lizard could have its attention diverted from its prey to the severed tail by this action. Even so, lizards still fall victims in large numbers to snakes, birds and other predators.

This species is the furze evvet of the New Forest, and the harriman of Shropshire. In Cheshire it is the swift. In suitable situations, such as sandhills, fallows, heaths and moors, it may be found all over Great Britain, including the Isle of Man; in most localities it is common. It is the one true reptile that Ireland possesses, and it appears to occur in all parts of the island, though not in any abundance. Outside Britain it is found across Europe and Asia, except for southern and south-east Asia.

The Sand Lizard

Family LACERTIDAE   *Lacerta agilis*

The sand lizard is of heavier build than the common lizard, and its legs are slightly shorter. The two are, however, sufficiently alike at first glance to make an error in identification possible, and some of the earlier records of the occurrence of the sand lizard have had to be severely revised as a consequence. In Britain the sand lizard is found only in certain restricted localities on the sandy heathlands of the southern counties of Dorset, Hampshire and Surrey, and the sandhills by the sea in Lancashire and Cheshire. Its southern habitats agree closely with those of the smooth snake, and where their ranges coincide the lizard provides a favourite food of the snake. The sand lizard is not found in either Scotland or Ireland.

The maximum size recorded for an adult male is just over $7\frac{1}{2}$ in. (190 mm.) long, of which well over half is tail. The maximum for a female is slightly less. The usual colour is grey or light brown above, with three longitudinal series of dark brown or black, irregularly shaped spots, each with a white centre, the one series along the mid-line of the back, the other two along the sides. The spots along the back may be edged with white, and they sometimes run together to form a continuous stripe. The flanks of the female are purple-brown, those of the male showing a marked tendency to a green suffusion. There is, however, a fair amount of variation in the colour, as is usual in lizards. The underparts are whitish or cream, with black spots, but the spots may sometimes be lacking. The green of the male becomes more pronounced during the breeding season, when the normally black-dotted yellow of the underside may also show some green.

The sand lizard burrows readily in the sand, but it will also use old mouse or vole burrows. It is more timid than the common lizard, and in spite of its trivial name, *agilis*, it is less agile and less able to climb than the second species. Its main foods are insects and spiders, and it will also take centipedes, woodlice, worms and slugs. Large prey is usually well shaken to be stunned or killed before being eaten.

Mating takes place in May and June, after the animals have come out of hibernation, which begins in late September or early October. During the breeding season there is a good deal of fighting between the males, the fighting attitude recalling that of cats, with the head lowered, neck puffed out, back arched and all four legs stiff. The female lays her six to thirteen eggs, the usual number being eight, in a shallow pit she digs herself. This is usually well concealed, covered with sand or leaves, or deposited under a stone. The older the female the larger the number of eggs she lays. The eggs have white shells of the consistency of parchment. They are laid in July, and the young hatch in the same month or early in August. The young sand lizards are grey-brown above and whitish below. Sexual maturity is reached at twenty-one to twenty-two months; full size is attained at four to five years of age.

Like the common lizard, the sand lizard is very apt to lose its tail by autotomy (see p. 177), short-tailed individuals often being seen, the original tail having been shed and another grown.

There are two species of lizards native in the Channel Islands, and by an anomaly, only one of these is usually included in lists of British animals, but even then only because the islands are politically British. The fauna and flora of the Channel Islands are comparable to those of the nearest mainland, France, and therefore none of the species should be included on British lists unless they occur also in England, Scotland, Wales or Ireland. The two species referred to are the green lizard (*Lacerta viridis*), with tail equal to three-quarters of its entire length, and the wall lizard (*L. muralis*) of variable brown colouration and a tail one and a half times the length of the head and body. The green lizard may sometimes be seen in this country as an escaper from captivity, being a favourite subject with those who keep vivaria.

The Slow-worm

Family ANGUIDAE   *Anguis fragilis*

The snake-like slow-worm, alternatively known as the blind-worm, or dead-adder, is a legless lizard. Internally, there are vestiges of the shoulder- and hip-girdles, evidence that its ancestors moved on four legs. A slow-worm also has eyelids like other lizards, the two halves of its lower jaw are joined in front, another characteristic of the lizards, and its tongue is notched, not forked like that of a snake.

The slow-worm may attain a maximum length of 17–18 in. (430–460 mm.), but the average 'large' example is about 1 ft. (300 mm.) long. The head is small and short, not so broad as the body immediately behind it. The tail, which is much longer than the head and body combined and longer in the male than in the female, tapers gradually to end in a sharp point. Often the tapering is absent, because the tail has been thrown off. Slow-worms part more readily with the tail than either of our two other lizards, but the new tail is never as perfect as the part it replaces. There is usually a ragged end to the old part, and the narrower new part appears as if clumsily thrust inside the fringe of old scales.

The scales covering both upper and lower sides of the body are nearly uniform in size and shape, but there are two rows in the mid-line of both back and belly which are slightly broader than the rest. Typically, all scales are broader than in the other lizards and rounded on the hindmargin, which is thinner than the dark coloured central part of the scale. Handling a live slow-worm gives a clear idea of its smoothness and the close attachment of its scaly covering. It feels, in fact, as if it has no scales. There is often a thin dark line down the centre of the back (usually indicating a female), and another on the upper part of each flank. The mouth is small and the jaws carry teeth uniform in size and slightly curved, so that the points are all directed backwards. The eyes are placed low down on the head. The head regions are not so clearly mapped out as in the other species of lizards.

The habitat is open woodlands, commons and heathland. They are most commonly seen in day-time in early spring sunning themselves, apparently lifeless and, because of their highly polished appearance, looking very like an animal cast in metal. It is at this time that they come into the open for mating and that, later, the pregnant females bask in the sun. They quickly make their way into the leaf-litter if disturbed. Later in the year they are more likely to be seen at dusk, coming out to feed. Slow-worms are regular burrowers, spending much time underground or lying in the earth with only the head showing. They can swim if necessary.

Knowledge of their feeding habits has largely been gained from captive individuals. Food preferences are clearly governed by the small size of the mouth. Spiders, small earthworms and tiny insects are taken and there is a marked preference for the small white slug (*Agriolimax agrestis*), so often a pest on tender green vegetables. This they consume in quantity, but where this slug is missing they will take others. The prey is seized in the middle and chewed from end to end. They will also eat snails, slowly pulling them from their shells as they swallow them. The principal feeding time is soon after sunset, or after rain, when the slugs themselves come out to feed.

Mating takes place from late April to June, the breeding season being marked by a great deal of fighting between the males, in which they try to seize each other by the head or neck. Once a hold has been obtained there is much writhing and rolling over together. In mating the male seizes the female by the neck and twines his body around hers. The female is ovo-viviparous, the eggs hatching within the body. The young are enclosed within a membranous envelope, which is punctured by a feebly developed egg-tooth, either at the moment of birth or shortly afterwards. Litters of six to twelve, but as few as four or as many as nineteen young have been recorded. They are born in late August or September, but if the weather is cold this may be delayed until October or later. The young are up to 3½ in. (90 mm.) long, silver or golden in colour with black underparts and a thin black line running down the middle of the back. Very active, they are able to fend for themselves from the moment of birth, catching insects, but showing a marked preference for slugs small enough to pass through their tiny mouths. On rare occasions the eggs are deposited before hatching.

Slow-worms spend the day-time under flat stones or logs and in burrows. Their principal natural enemies are the adder and the hedge-hog, but they are also taken by kestrels, buzzards, little owls and other birds of prey, smooth snakes, badgers, rats, foxes and poultry. Young slow-worms are particularly vulnerable and their enemies include frogs and toads. In October the slow-worm hibernates in an underground burrow, in a hollow beneath a large stone, or even beneath a pile of dead leaves. It is the first of our reptiles to reappear, at the very beginning of spring. As many as twenty may be found in one hibernaculum, the largest being underneath, the smallest on top. Slow-worms cast their skins or, more correctly, their cuticle about four times a year, but the frequency of the sloughing depends upon whether or not it is a good slug year, the shedding of the cuticle being in response to the need for more space for the growing body. The skin is shed whole as in snakes.

Slow-worms have been known to live in captivity for up to thirty years or more, the record being held by one that lived in the Copenhagen Museum for fifty-four years. At one year old a young slow-worm has doubled its length, to 6–7 in. (150–180 mm.); at two years old it is 8½–9 in. (about 220 mm.). Sexual maturity is attained at four to five years of age in the females and three years in the males, when the slow-worm is 11 in.–1 ft. (280–300 mm.) long.

It was in the slow-worm that the discovery was made in 1886 of vestiges of a degenerate median eye connected with the pineal gland. This find led to a good deal of investigation which, it was hoped, would reveal the function of this gland and the significance of the third, or pineal, eye. Apart from learning that the pineal body, in some form or other, is common to all vertebrates, and that the median eye in varying states of near perfection is present in several living reptiles and must have been present in a large number of extinct forms, we are still little the wiser on these two points.

The slow-worm is generally distributed throughout the British Isles, with the exception of Ireland. It is much more plentiful in the south and south-west of England than in the east or north, but even in the south it is much more abundant in some districts than in others. In southern England a variety is sometimes seen, known as the blue-spotted slow-worm (var. *colchica*). These are always males and the colour, varying from a light blue to deep ultramarine, may be present in spots or stripes, sometimes so closely set that the animal appears blue all over.

(*above*)
Common
frog

PLATE 37

(*right*)
Female
edible frog

(*above*) Palmate newt

PLATE 38

(*left*) Smooth newt larva

(*below*) Marsh frog

SNAKES

The Grass Snake

Family COLUBRIDAE   *Natrix natrix*

Before entering upon a description of the grass snake, it would be as well to say a few words on the structure of snakes, for in a general way our three native species are alike.

Snakes have no breast-bone, shoulder-blade or collar-bone. The ends of the ribs away from the backbone are free. As a result, when bulky food is taken the ribs can be pressed outwards and upwards to allow the necessary distension of the body. The tooth-bearing bones of the skull are connected loosely, and the head can be flattened and widened, so that the mouth can admit prey equal to three times the size of the snake's head, under normal conditions. One feature making this possible is that the angle of the mouth extends far behind the level of the eye. The jaws, also, move independently, each half of both upper and lower jaws being capable of movement on their own. The teeth are all pointed backwards which makes it difficult for living prey to struggle forward and escape once it has been seized. In engulfing the prey, the upper and lower jaws of each side are moved together, and movement on one side of the mouth alternates with that on the other side, the prey being dragged backwards, as it were, by a see-saw action. The teeth are not planted in sockets. If they should get broken, they are soon replaced by others which, lying in reserve on the inner side of the gum, move up into position.

The eyes of a snake are always wide open, for there are no movable eyelids to close them, and the eyeball has only slight power of movement under the brille or spectacle, as its transparent cover is called. There is no external ear, no ear-drum, tympanic cavity or Eustachian tube, and the ear-bone is attached to the jaw. It seems, therefore, that snakes do not hear in the ordinary sense. Experiments have shown that they do not react to noises conveyed through the air, but will respond to such things as footsteps, presumably as a result of vibrations through the ground picked up by the chin and transmitted through the bones of the skull. The tongue is long and forked, and it is constantly used when a snake is active, being protruded through a small gap in the front of the upper jaw. Snakes probably have little sense of taste, but they have a good sense of smell. In addition to the nostrils they have a pair of Jacobson's organs These lie one on either side of the roof of the mouth between the palate and the nasal cavities. Each has a short duct opening in the roof of the mouth. As the tongue is thrust out it picks up particles of scent. When it is withdrawn the tips of its two branches are thrust up into the Jacobson's organs. In this way snakes pick up the trails of their prey, and also locate the trails of other snakes when seeking a mate or making their way to a hibernaculum.

The grass snake is our largest British species, fully grown females averaging 4 ft. (1·2 m.) in length, the males 1 ft. (30 cm.) less. In this country there is a record of a female 5 ft. 9 in. (1·75 m.) long. Such a

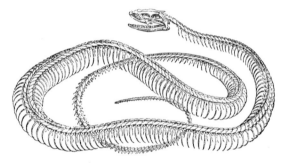

Skeleton of snake

snake looks truly formidable. Larger examples are found in southern Europe, where the record is 6 ft. 8 in. (2 m.). This snake is of graceful form, the body tapering gently from its middle to the very slender tip of the tail. The ground colour is olive-grey, olive-brown or olive-green, and this is uniform to the tip of the tail. Along the back are two rows of small blackish spots, and there is a row of short vertical bars along each side. The underside, which is covered with broad plates, is chequered in black or grey, and white, but it is sometimes entirely black. The tail accounts for about one-fifth of the total length. The long, narrow head, covered with large shields, ends in a blunt snout, with the nostrils lying well to the sides. The upper lip is yellow or white. The rather large eyes have round pupils circled with gold and a dark brown iris, and just behind the head are two patches of yellow, or orange, pink or white, forming a bright collar, whence the name ringed snake, which serves to identify this species at a glance, except in large females in which this collar is sometimes missing. Immediately behind it are two patches of black, often united in the middle line. These and the size, general colouration and distinctive patterning (particularly the black bars along the sides of the body) make this harmless species easy to distinguish from the adder, which is smaller, darker, has a black zig-zag line along the back and a black V on the head. Nevertheless, every year large numbers of grass snakes are killed in the belief that they are poisonous.

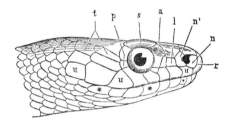

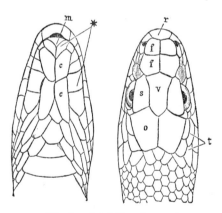

Apart from the head-shields and the broad plates of the underside, the grass snake is covered with nineteen longitudinal rows of small, overlapping, lance-shaped scales each with a central ridge or keel. These scales are an outgrowth from the skin, and when the snake moults they do not fall off as the hairs of fur-clad animals do, but the entire skin with its scales is usually cast intact, although it may sometimes be torn into pieces. The old skin separates first at the edges of the jaws, and the snake pushes

The head shields of a snake

r rostrol shield; f anterior and posterior frontal; v interparietal; s supraocular; o parietal; n nasal; l loreal; a preocular; p postocular; u upper labial; t temporal; m mental; * lower labial; c chin-shield.   After Günther against the ground, or against stones or the stems of plants until the loose skin has been pushed behind the head. Then the snake glides out of the remainder, reversing it in the process. In these discarded sloughs the spectacle-like covering of the eye will be found unbroken.

Although the grass snake may be found frequently near ponds and ditches it is by no means restricted to such places; it may be met with on chalk hills, sandy heaths, open woodlands, under hedgerows and in marshy places, often far removed from water. Its main item of food is frogs, but it will also take toads and newts; and it feeds occasionally on fish, mice and small birds, as well as lizards, voles and birds' eggs. The young grass snake takes worms, slugs, tadpoles, as well as young newts, frogs and toads. It swims well and often enters the water to obtain its prey. It can also climb well. Although an agile reptile, it may be caught without difficulty where the ground is not too rich in mouse runs or too well covered with furze which provide a ready refuge. The undulations by which it progresses are always horizontal, not vertical. When captured it has the habit of giving off a foul smell with a strong odour of garlic among the other objectionable scents, but it seldom makes any attempt at biting, though it will hiss freely and snap its jaws. The third line of defence is to feign death. The snake turns over on to its back, becomes rigid, its mouth is opened and the tongue lolls out. To all intents and purposes, the snake is as dead as a doornail and will show no signs of life if picked up and handled. Moreover, if it is then laid down on the ground the right way up it will turn over on to its back, go rigid once again, with the tongue lolling. A snake handled will use these stratagems, usually in the order in which they are described here, although it may not necessarily go through all the three stages. In spite of this a grass snake can soon become tame.

In the autumn the grass snake retires to a shelter under the roots of trees, in banks or hedge-bottom, or under a brushwood pile for its winter's sleep. Usually a number of them assemble in the same hibernaculum, their bodies intertwined. They remain there until March or April, when the frogs, toads and newts, having already emerged from a similar retirement, are available for food. During April and May, mating takes place. This is usually preceded by a courtship in which the male approaches the female with nodding movements of the head. Then, he lays his chin on her back and slides it forward towards her head. The tongue is in motion throughout this performance. Finally, he seizes the female by the back of the neck in his jaws, and with their bodies entwined pairing takes place. Some time between June and August the female seeks some convenient mass of fermenting vegetable matter into which to burrow and deposit her eggs. If a heap of fresh stable manure is available she will use it, the heat hastening incubation, but a number of other materials has been used. Heaps of decaying leaves, piles of sawdust in sawmills, or hayricks are used, but if none of these is available an old tree trunk or even ploughed earth will suffice. The eggs, which may number a dozen or anything up to four dozen, are equal-ended ovals with a tough, parchment-like shell, and are all connected to make a string. They are laid about two months after pairing has taken place, the time occupied by the laying being ten to twelve hours. The female burrows into the heap with her snout and enlarges the cavity by contortions of her body. Sometimes several females will use the same nest and as many as 250 eggs have been found in one place.

As the eggs are laid they are covered with a secretion from the oviducts which sticks them together. Later, they begin to absorb moisture from their surroundings, and increase in size until they are about $1\frac{1}{4}$ in. (30 mm.) in length. They hatch in six to ten weeks, according to temperature, and it has been recorded occasionally that the female coils herself around the eggs and appears to resent interference. The young grass snakes, which measure $6-7\frac{1}{2}$ in. (150–190 mm.), make their way out of the egg by tearing several rents in it with a special egg-tooth projecting from the front of the jaws. Within a few hours of hatching the tooth becomes loose and drops off, its special function having been performed. The young snake sheds its skin before taking its first meal, and thereafter goes through the same process four or five times in a year.

The grass snake appears to live for at least ten years, judging from marked specimens measured in successive years. The female is about four years old with a length of 2 ft. (60 cm.), before she begins to breed, but the male is sexually mature at three years.

The grass snake is widely distributed over England and Wales, and although it has been recorded in the south-eastern parts of Scotland, these records are not strongly credited. Nevertheless, I have had several authoritative records of grass snakes being picked up as far north as Inverness. It appears, however, never to have reached Ireland.

The Smooth Snake

Family COLUBRIDAE   *Coronella austriaca*

Although in general features it is similar to the grass snake, the smooth snake has something of the appearance of an adder to the casual observer. This is largely due to the series of black or dark brown spots over the length of the back. These may be in pairs or they may form crossbars. In any event, a closer look soon shows that the markings are very different from the zig-zag on the adder. In the hand, the smooth snake exhibits a sufficient number of differences to make its identification easy. The smoothness which has resulted in its name is at once evident to the touch, and is due to the fact that all its scales lack the keels or ridges. The longitudinal rows of small scales on the back and sides of both of our non-venomous snakes are nineteen in number. In the viper there are twenty-one rows, rarely nineteen or twenty-three. Each one of these scales is marked with a tiny pit which appears to coincide with the end of a nerve fibre, so that it would appear that the sense of touch exists in every separate scale.

The smooth snake never attains to such a great size as the grass snake, its maximum length being 2 ft. 1 in. (630 mm.) for a female, and just over 22 in. (560 mm.) for a male. In Britain it rarely exceeds 18 in. (460 mm.). Its head is not so distinctly marked from the body as in the grass snake, and the slender tail is a quarter of the entire length in the male and one-sixth in the female. The ground colour on the upper side is grey, brown or reddish, with small black, brown or red spots usually in pairs or crossbars, already mentioned. Occasionally, another less distinct longitudinal series of spots runs along the sides. The upper part of the head is mainly brown with a black patch at the back of the head; in young individuals more especially it is blackish throughout. A dark streak runs from the nostrils and through the eye to the angle of the mouth. This streak may be carried along the sides, as far as the tail. On the underside the colouring is some tint of orange, red, brown, grey or black, either uniform or with white spots or dots. The sides of the belly are usually whitish. The eye has a round pupil, as in the grass snake, which helps to make it look less sinister than the adder.

This species has been found in the New Forest and other parts of Hampshire, in Dorset, Surrey and Berkshire. Wiltshire also has yielded some specimens. There have also been records for Kent, Somerset and Devon, but these are believed to be based upon faulty identifications. The smooth snake may be abundant locally, especially where the sand lizard occurs, this being the smooth snake's main prey. The usual habitat includes heaths, stony wastes and wooded hillsides, especially near water into which the snake readily slides when alarmed, hiding itself in the mud at the bottom. The smooth snake is much given to burrowing and spends a fair proportion of its time underground. Its favourite food is lizards, but it also takes young snakes and slow-worms, and occasionally mice, voles and shrews. When these last are sufficiently large the snake will coil itself around them in boa-constrictor fashion, but this is merely to hold them while being swallowed.

Pairing takes place soon after the emergence from hibernation in spring. As in the case of the slow-worm and the common lizard, the eggs are retained until the young are ready to hatch out, and they are born about the end of August or in September, but cold weather may delay this until October. The litters vary in size from two to fifteen, but usually there are about six to a birth. They are enveloped in a thin membrane which is ruptured immediately it leaves the parent body, and the young snakes are then about 5–6 in. (130–150 mm.) in length.

Like the grass snake, this species emits an objectionable odour when captured, and at first attempts to bite. This phase soon passes, however, and it becomes more tame than our other species. This tameness and its readiness to co-operate give the smooth snake an air of intelligence which the other two species do not possess.

The smooth snake is found throughout the greater part of Europe. It was not discovered in Britain until 1853, when F. Bond found one in the New Forest, but it was left to J. E. Gray to give the first scientific record of it in 1859. The publication of this record set people looking for it. Then it was discovered to be very numerous in several restricted localities, notably near Bournemouth.

The Adder or Viper

Family VIPERIDAE   *Vipera berus*

In contrast to the long gracefully tapered body of our other two snakes, the adder or viper is short and thick in the body and has a short tail. The length of an average male is about 21 in. (530 mm.), the average female being 2 ft. (600 mm.). The record length for an adder is 2 ft. 8 in. (810 mm.). The head is flatter above, and broadens behind the eyes, so that it is quite distinct from the body. Further, the shields on the head are very much smaller than the corresponding plates of the grass snake. The iris of the eye is coppery-red, and the pupil is a vertical ellipse, a feature which usually denotes nocturnal habits, although adders are also active by day. To what extent they are nocturnal or diurnal is not known for certain. It is certain that adders are given to basking, but so are other nocturnal or mainly nocturnal animals. The point about the adder's eye is that it is rich in rods, which are markedly more sensitive to light than the cones. An animal with an eye rich in rods will see well at night, but during the day needs protection from intense light. The split pupil cuts down the intensity of the light. So it may be that adders are as much nocturnal as diurnal.

There is considerable variation in both the colour and the markings of adders and this is one of the few snakes in which male and female are coloured differently. Generally speaking the ground colour may be said to be some tint of brown, olive, grey or cream, but it may be so dark that the darker markings are scarcely perceptible at first sight. Along the sides there are whitish spots, sometimes reduced to mere dots. Those adders that are cream, dirty-yellow, silvery-grey, pale grey or light olive with jet black markings are usually males. Red, reddish-brown or golden individuals, with the darker red or brown patterns, are females. The throat of the male is black, or whitish with scales spotted or edged with black. In females the yellowish-white chin and throat are sometimes tinged with red. The eyes of the female are smaller than those of the male.

The most characteristic mark is the dark zig-zag line down the centre of the back, with a series of spots on either side. This may be broken up into oval spots or its indentations may be filled in to form a continuous stripe. The other characteristic is a pair of dark bars on the head, and these may form either a V or an X. The broad shields which cover the lower surface may be grey, brown, bluish, black, or bluish with triangular spots of black, sometimes with white dots along the margins. Below the end of the tail the colour is yellow or orange. There are records of specimens that are almost entirely rich black, the excepted portion being the whitish underside of the head and throat.

Adders are usually found in dry places, such as sandy heaths, dry moors, the sunny slopes of hills and hedgebanks; also bramble clumps, nettle beds, heaps of stones and sunny places in the woods. Sometimes, however, they live in heathy and grassy places that are damp or permanently wet. They feed mainly on lizards, but also take small mammals such as mice, shrews and voles, young weasels, as well as birds, slow-worms, frogs, newts and large slugs. The young subsist at first on insects and worms. Adders kill with their poison fangs, striking with a swift thrust of the head. The victim dies within thirty seconds or so, but in that time will escape to cover. After waiting for a while the snake follows, seeking its victim by sight and by the aid of the flickering tongue. Frogs and newts are often eaten straight away, without being killed first.

The adder retires in autumn to a hollow under dry moss among the heather, under heaps of brushwood or into discarded, leaf-covered ground nests of birds. It reappears about April, and may then be seen coiled on a sunny bank apparently more concerned with absorbing heat than finding food. It may emerge as early as February if the weather is warm, or as late as May in a cold spring. Adders pair in April or May, but ovulation does not occur until the end of May. The young, varying from five to twenty, are born in August or September. This species is also ovo-viviparous, the eggs being retained until fully developed, and when the young are born they are often coiled up tightly in a thin, transparent membrane, which is usually broken during the process of birth. The young adder possesses an egg-tooth, which is not shed for some days after birth, but it is so situated that it does not appear to be capable of use for rupturing the membrane, which is achieved by convulsive movements of the body and head of the baby snake followed by a forward thrust of its snout. The newly born young measure from 6–8 in. (150–200 mm.) at birth, and are immediately capable of an independent existence.

The very old legend that young snakes when alarmed seek refuge in the mother's stomach seems to be founded upon their habit of sheltering under her body on such occasions. This matter has been very fully investigated in recent years, and it is beyond reasonable doubt that it is a physical impossibility for the young to use the mother's mouth and gullet as a bolt-hole.

At the start of the breeding season there is a good deal of rivalry between the males. One expression of this is a dance which has been mistaken for courtship. In this, two males face each other with the head and fore part of the body held erect, the hind part of the body coiled on the ground. They sway from side to side, then, with the fore parts of their bodies intertwined, they push and thrust each other, each trying to force the other to the ground. Eventually one gives up and departs. The behaviour of a male towards a female is similar to that described for the mating of other snakes. He approaches her and lays his chin on her back, near her tail, and slides his chin forward until it is over her neck, tapping with the chin as he goes and continually clicking his tongue in and out. Finally, he takes her neck in his jaws, throws his body in a loop over hers and mating takes place.

The adder is not so amenable to a life of captivity as our other snakes. It is apt to refuse all food and most captive adders die of starvation. On being captured they are ready to bite, as they are timid and readily alarmed by sudden movements. In a state of freedom they are not as dangerous as is popularly supposed. They are more concerned with getting under cover than with striking. Accidents from their bites are rare in this country, where people go about well shod, so there are very few authenticated cases of death from adder-bite. Where deaths have occurred they have been usually in young children or elderly people. The toes or the fingers are the most likely to be bitten, for the adder's

mouth is not large enough to enable it to bite the larger parts. In case
of a bite one should remain calm, as panic does more damage than
anything. The standard first-aid treatment is to apply a ligature above
the wound, to prevent the poison spreading, but the ligature must be
loosened periodically to allow the circulation to continue. Meanwhile,
the bite can be sucked with the mouth, provided there are no abrasions,
but medical aid should be sought as soon as possible, and it is doubtful
whether sucking the wound is desirable or necessary with adder-bite.
Some authorities recommend cold compresses on the wound as the best
first-aid treatment. The symptoms of poisoning vary from slight to
severe, according to the circumstances. They include giddiness, vomiting,
diarrhoea and prostration, sometimes with loss of consciousness which
may not necessarily persist. In fatal cases there is a weakening of the
pulse and breathing, and death may ensue in six to sixty hours after
the incident. During the first half of this century there was an average
of one case of adder-bite a year, of which seven proved fatal, although
anxiety was felt by those in charge in a score or so of the others. There
are, however, probably many minor cases that never reach the doctor's
notice.

The adder is found in all parts of Britain, but seems to be progressively
less common from the Thames northwards. It is not known in Ireland.

AMPHIBIANS

The Crested Newt

Family SALAMANDRIDAE   *Triturus cristatus*

Newts, of which there are three British species, are generally similar to frogs and toads in their mode of life and in having an aquatic tadpole, but they differ in retaining throughout life the compressed tail. As the structure, development and habits of the three species are much alike, a general description is given here before dealing with each separately.

The body is elongated. The two pairs of legs are almost of the same length, the hinder pair being slightly longer. The hands have four fingers and the feet five toes as have the other amphibians. The skin is

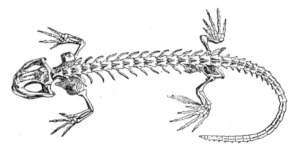

Skeleton of newt

without scales, and equipped with a mucous and sensory apparatus, arranged principally along the flanks and the base of the tail. On the head are two parallel rows of dark coloured pores, with others scattered on the sides of the head, and another series extending along the flank and tail on each side. These are sensory cells, in small depressions in the skin, corresponding to the lateral line in fishes, functioning as a mechanism for posture and for detecting vibrations in the water. The cells are present in the tadpole and adult, but atrophy partially during the period spent on land. During the breeding season the skin of the males grows up into a crest or fin along the middle of the back and above and below the tail, and the toes are more or less broadly fringed on each side. These outgrowths of skin are sexual adornments, as well as an aid to swimming, the male being the more active partner at this time. The crests are also rich in sensory organs.

Newts are terrestrial outside the breeding season, but return to water to breed, finding the ponds by haphazard methods rather than deliberate migration, the urge to do so being due to hormonal secretion.

The skin serves the same function of respiration as in the frog (q.v.), and like the frog, newts are compelled when on land to force air into their lungs by a constant pumping and swallowing action of the mouth and throat.

In the breeding season the male stimulates the female to breeding condition by displaying his crest and his heightened colours; also by butting her with his head and lashing with his tail. In this tail-beating,

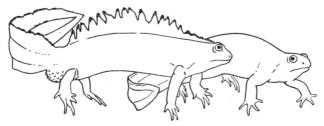

Male crested newt, tail-beating when courting

as it is called, the tail does not strike the female flank, but appears to be used to set up vibrations in the water which doubtless impinge on her sensory lateral line and exert an excitory effect. A similar action is seen in the courtship of many fishes. He further stimulates the female by secretions from hedonic glands on his cheeks, body and cloaca. Often the male will position himself in front of and at right angles to the female with his tail bent towards her and vibrating (tail-beating). At the end of the courtship, the male emits a spermatophore in the form of a mushroom-shaped gelatinous mass, the head of which is packed with sperms. The spermatophore sinks to the bottom, the female moves forward until she is in position with the cloaca immediately over the spermatophore, which is then taken through the cloaca as her body is pressed on to it. The eggs are, therefore, fertilised internally. They are laid one at a time against a long leaf of one of the pond-plants which is then folded over by the female, who uses her hind-feet almost as hands, and adheres to the egg.

The eggs hatch in three weeks or so, the tadpoles freeing themselves by digesting the egg capsule by means of special glands on the snout. The liberated larvae are more slender and fish-like than the tadpoles of the frog. They have three pairs of external gills, and soon after hatching they develop two pairs of stalk-like organs with swollen ends from the sides of the upper jaw, which enable them to cling to water plants. Development is more prolonged than in the frogs and toads, but it is mostly complete at the end of summer before hibernation begins. The young newts then crawl out of the water and seek shelter under stones in the immediate neighbourhood of the pond. Sexual maturity is attained in three to four years.

Newts have many enemies. The tadpoles are eaten by sticklebacks and aquatic insects. Adults are eaten by fishes, snakes, water birds and, on land, by hedgehogs, stoats, weasels and rats. There is an unpleasant secretion from the skin of the adult, but this seems to have little protective value against predators. Newts can regenerate lost or damaged limbs, the young being capable of doing so more readily than the adults.

The crested newt, also called warty or great newt, is our largest species, attaining a maximum length of 6 in. (150 mm.), of which 2½ in. (64 mm.) are tail. The skin on the upper parts is dark grey or blackish-brown and is covered with small warts. Along the lower flanks is a sprinkling of white dots and the underside is coloured yellow or orange, boldly spotted or blotched with black. There is a strong collar-like fold at the base of the throat. The male's nuptial crest starts from the head as a low frill, but between the shoulders and the thighs becomes high with its edge deeply notched, the resulting 'teeth' waving freely in the water. Behind the thighs there is a gap, and then the crest rises again as a tail fin, the lower edge of the tail having a similar extension of skin. After the breeding season the crest is absorbed, leaving only a low ridge. Along the sides of the tail proper runs a bluish-white, silvery stripe. The eye has a golden-yellow iris.

The female is larger than the male and similar in colour except that

the lower edge of her tail is yellow or orange. Above the spine runs a
fine depression, the whole length of the body. This is permanent but is
coloured yellow in the breeding season, which begins in April.

Warty newts usually hibernate on land, in cracks in the ground, under
stones or logs or in dense grass litter. In early March they return to the
water. Egg-laying begins in early April, each female laying 200–300 eggs.
This may continue until the middle of July, when the adults return to
land. Some delay their departure from the water until August. The
newly hatched semi-transparent larvae are yellowish-green with two
black stripes along the back which, later, when the ground colour
changes to olive, become broken up into spots, and the flanks and the
underside become tinged with gold. They have a finer set of plumed
gills than the frog tadpoles, and their form is more graceful and not
'big headed'. They also breathe through the skin of the whole body.
Some individuals do not complete their development before winter, and
remain in the pond until spring. They may be frozen in solid ice, but
they thaw out none the worse for this. Their food, detected by sight and,
especially in water by smell, consists of any small aquatic life such as
insects, worms, crustaceans, frog-spawn and tadpoles. On land they
feed upon worms, slugs, snails and insects. There are minute teeth along
the jaws and on the palate, but they serve only for the retention of living
food. The tongue is attached to the front of the mouth and is used,
especially by the young, for capturing prey.

Sometimes adults which do not leave the water immediately after the
breeding season, may come on land in the autumn, assembling in
numbers in a damp hole, where they twist and intertwine into a ball,
apparently by this means preventing loss of moisture. In this way they
pass the winter in an almost torpid condition. Alternatively, they may
pass the winter in the water.

Newts shed their skin much in the same manner as a snake, separation
beginning at the lips and by the help of the hands and wrigglings of the
body the skin is worked back over the tail. These sloughs may be found
floating entire in the water looking like newt-ghosts; but on land they
may be disposed of piecemeal, the old skin being sometimes swallowed,
as in the toad.

The crested newt which is widely but locally distributed over England, is less plentiful in the west. It is present in Wales and Scotland, but in both there are large areas in which it does not occur. It is absent from Ireland. It is also the largest European newt.

The Smooth Newt

Family SALAMANDRIDAE   *Triturus vulgaris*

The smooth newt, common newt, spotted newt, eft or evat is the best known of our three species. It is widespread throughout England and widespread but uncommon in Wales; sparsely distributed in Scotland and Ireland. It is very much smaller than the crested newt, its maximum length being under 4 in. (101 mm.). It varies in colour, but the prevailing tint is olive-brown with darker spots over the upper side, and dark streaks on the head. The underside is orange or vermilion with round black spots, the colours becoming more intense in the breeding season. The throat is yellow or white, mostly dotted with black. The underside of the female is, as a rule, much paler than that of the male, and often unspotted. The eye has a golden iris. The female has shorter fingers and toes than the male. At the breeding season the male develops a continuous crest, the upper edge of which is festooned instead of being serrated, running from the top of the head to the end of the tail. The lower edge of the tail has a spotted blue band with black base. The toes of the hind-feet in the male develop fringes which disappear at the end of the breeding season.

The breeding habits of the smooth newt are much the same as those of the crested newt. Hibernation is on land, and in March or early April a return to water is made. The courtship lasts through April, but both male and female remain in the water until July or later. Each female lays between 200 and 350 eggs. The larva, light green or brownish above, is spotted with yellow along the sides and tail; the tail ends in a thread-like prolongation of its tip. Its change to the adult, which includes breathing by lungs instead of gills, is completed in 15–17 weeks. The smooth newt becomes sexually mature at three years of age.

A state of partial neoteny is sometimes found in this species. In complete neoteny the larva fails to metamorphose and may become

(*above*) Grey or Atlantic
seals

PLATE 39

(*right*) Common seal
(Note strong develop-
ment of whiskers)

(*above*)
Natterjack
toad

PLATE 40

(*left*)
Defence
posture of
common toad
when con-
fronted by a
grass snake

sexually mature while still having a larval form. The olm, of Europe, is an example. Partial neoteny means that the metamorphosis is delayed, and the larva may increase in size but the sex organs do not develop. This is not to be confused with over-wintering when the larva fails to metamorphose before the winter and hibernates as a tadpole.

After the breeding season, in July, the adults and young leave the water and seek their food at night among the vegetation on land. For the rest of the summer, and in the autumn and winter, they hide under stones or logs, in cracks in the earth or in dense grass litter. They become duller in colour, and the skin becomes more opaque with a fine velvety surface. They are then known locally as dry evats.

The Palmate Newt

Family SALAMANDRIDAE   *Triturus helveticus*

In general appearance the palmate newt is similar to the smooth newt and it is as smooth. It was not recognised as a distinct species until 1787, when it was discovered in Switzerland, and it was not recorded in Britain until 1843, when it was found at Bridgwater, in Somerset. Until half a century ago it was still considered rare, but closer examination shows that while it is local in the south-east of England, it is more plentiful than the smooth newt in the west.

It is smaller than the smooth newt, its length being 3 in. (76 mm.) only. In the breeding season its distinctness is evident, for the male has then a nearly four-sided body in cross-section owing to the development of a ridge of skin along each side of the back. All three British newts have a glandular pad on each side of the body, which begins on the neck and ends at the base of the tail. It is the upper margin of this that forms the ridge in the palmate newt. The dorsal crest, instead of being high in front and having an undulating edge, rises gradually from the head, is of less height than in the other two species and has a straight non-convoluted margin. The tail appears to have had the tip cut off and the attempt to renew it to have resulted only in the development of a short thread from the centre of the cut portion. But the species gets its name from the black web which connects the toes of the hind-foot in

15

the male during the breeding season. The tail develops a fin along its lower edge in both sexes, and this in the male is edged with blue and in the female with orange. Another point of distinction lies in the colour of the throat. Instead of the black-dotted white or yellow of the smooth newt, the throat of the palmate newt is flesh-coloured without dots.

Above, the colour is olive-brown with darker spots; the underparts are orange bordered by pale yellow, with or without black spots.

The tadpoles resemble those of the smooth newt, and the number of eggs laid by the female may be from 300–400.

After the breeding season, the webbing of the feet of the male becomes reduced to a margin along each toe and no longer constitutes a palm, but the truncated tail remains, as a specific distinction, though the thread-like prolongation becomes very much shorter, to the same length as in the female.

The palmate newt lives more on high ground, although it may occur at sea-level, often in company with the two other newts. At the lower altitudes it leaves the water at the end of the breeding season, but on higher ground it does not forsake the water for long periods at any time of the year. Its distribution in Great Britain follows the higher configuration of the country, so that it is absent over a wide area of the east Midlands and East Anglia. It is not found in Ireland, and outside Britain its range includes only western Europe.

The Common Frog

Family RANIDAE   *Rana temporaria*

The common frog is extremely variable in colour. The ground colour of the upper parts may be grey, olive, yellow, brown, orange or red, and this may be speckled, spotted or marbled with brown, red or black. There are, however, certain markings consistently present. These include the dark cross-bars on the hind-legs, the dark streak on the fore-arm, the dark streak in front of the eye and a patch of brown behind each eye. The underparts are dirty-white to pale yellow in the males, yellow to orange in the females, in both speckled with grey or brown, or, in the females only, with orange or red. At least some of the variation in colour is due to the situation in which frogs are found. Red frogs, for example, are especially numerous in Scotland. In all, however, there is a marked tendency for the colour to change from light to dark according to the surroundings, the pigment cells of the skin being able to expand and contract under the influence of varying intensities of light reflected from the surroundings.

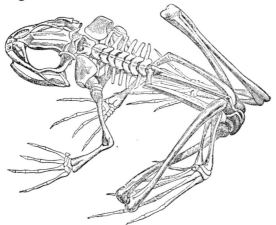

Skeleton of frog

The frog's fore-limbs are short compared with the hind pair, and the four moderate-sized fingers are without webs. The bones of the hind-limbs are unusually lengthened, and this is especially true of those of the ankle, giving the legs the appearance of having a supplementary joint. The hind-leg is, in fact, one and a half times the length of head and body combined. The hind-foot has five long toes connected for half their length by a web of skin, and is a most efficient paddle for use in the water. The horizontal gape of the mouth extends back beyond the eye. The prominent eyes are perched up on the forehead, and each has a fine golden iris, speckled with brown, and a horizontal pupil. A large circular depression behind and below the eye is the site of the ear-drum. The skin is smooth with numerous small smooth warts, especially on the flanks. These are minute mucous glands which keep the skin moist. Slightly larger glands form a pale line running back from the eye on either side. In the breeding season the skin of the male becomes more slimy and the warts on the skin of the female become larger and then tend to be pearly white.

Frogs differ from snakes but resemble lizards in having eyelids. Like birds, they have an additional lid, the nictitating membrane. There is a row of delicate teeth along the upper jaw, but none on the lower jaw, and there are other teeth on the palate. The deeply notched tongue is attached by its base to the front part of the mouth, the tip lying far in towards the throat. In use it is suddenly turned over and projected forwards so that for almost its whole length it is beyond the muzzle. It is always stated that the tongue is sticky, which poses the question of how an insect, caught on the tongue, becomes detached when the tongue is drawn back into the mouth. High-speed photography shows that the ends of the tongue whip round the prey, giving a temporary grip, and this must be assisted by the granular surface of the tongue.

A frog has no neck, the base of its skull lying close to the collar-bones, and there are only a few pairs of very short ribs. Being without the framework of ribs and associated muscles found in most air-breathing vertebrates, it fills its lungs by forcing air down with a pumping action of the throat, which can be seen externally in the movements of the throat.

The skin plays an important part in the oxygenation of the blood. It has been shown that a frog experimentally deprived of the use of its lungs can breathe for a long period through the skin alone. Respiration through the skin is, indeed, more important than the lung-breathing, even when on land; when hibernating under water it is the only means of respiration.

The female is stouter in build than the male, and she is relatively longer in the body and shorter in the leg. A distinguishing feature of the male is a pad on the first finger which in the breeding season becomes large, black and roughened, and assists him to hold his mate. This is shed at the end of the breeding season.

Careful records kept over the years show the spawning period of the common frog to extend from February 8th to nearly the end of March, varying with the season and locality; in general spawning is retarded when the weather is cold and is progressively later as we go north and east.

It is often asserted that 'any pool of water will do' for the spawning, but this is not strictly so. Frogs usually spawn in a few inches of water, but spawn may be found in water several feet deep. The ponds may be small or large, and the fact that small shallow ponds tend to dry up if there is a lack of rain, so that the spawn dries with them, seems to have encouraged the idea that any pool will do, however unsuitable. I have made special observation of this around my house, in Surrey, and find the same ponds visited each year. Certain ponds never contain spawn although others less than a hundred yards away from them do so consistently. An artificial pond in my garden is always used by both frogs and toads. Half a mile away, in the cart-track through a wood, frogs spawn every year in water that has settled in ruts made by wheels, but these are usually doomed.

Having a breeding pond so near home has made it possible to keep close watch on day-to-day events. One year I had the good chance to see a frog emerge from the ground, where presumably it had hibernated. The spot was about 50 yd. (45 m.) from the pond. The frog moved a yard or so uncertainly as to direction, then set off in a direct line to the pond in a series of leaps, at a speed which I would not previously have credited. It was the first to reach the pond, and during the next two days I watched frogs come in from all points of the compass, and most of them were seen arriving along a straight course, in a similar series of long hops. This was very obvious as they leapt through the air, above the vegetation, over the last 20 yd. (18 m.) or so. Nothing could illustrate more effectively the purposeful migration to a selected pond.

It has been suggested that the frog is guided to the pond by the scent from the essential oils of the filamentous alga upon which the tadpoles feed. It is even less certain what brings them out of hibernation. In 1955, for example, there was thick ice on all the ponds in my locality, and as soon as this had gone the frogs were in the ponds. Although the severe cold was ended we had several spells bringing thin ice on the ponds, but this did not seem to affect their breeding. Many theories on the subject have been propounded, e.g. warm wet nights favour the migration; the direction of the wind is of importance, etc., but it is usually easy to make observations which conform to none of the suggested rules. It is, in fact, a subject requiring much more investigation.

The mystery has deepened with the times. In recent years there have been several instances of ponds having been filled in and housing estates built on the land. In the following spring occupants of the new houses have been plagued by swarms of frogs at the breeding season. They could not have been guided there by the smell of an alga now gone or by land-marks now removed. The only reasonable assumption must be that they have made their way by celestial navigation, using the sun by day and the stars by night, like migrating birds.

Sometimes the frogs may assemble at a pond and spawning will not take place for a week or ten days. In other instances spawning will have been completed in two to three days after the arrival of the first females. The males are the first to arrive and most of them are there before the first females appear. It may happen, however, that towards the end of the breeding season some males and females, often in amplexus, can be seen arriving. Again, when all have mated and dispersed, the solitary frog may be found in the pond, lonely among the dense mass of spawn.

The common frog croaks only at the breeding season, although occasional croaks may be heard in autumn, just before hibernation. This is because they are already coming into breeding condition before they go into their winter quarters. The vocal sacs are internal, and when distended are visible only as a swelling on either side of the throat. The croak is produced by passing air backwards and forwards across the vocal chords, from the lungs to the mouth and back again. The mouth does not need to be open, therefore, and frogs can sometimes be heard croaking under water. A second vocalisation is the grunt uttered by a male when grasped by another male. It is a means of sex recognition. Another, infrequent, sound made is a scream, uttered with the mouth open, when a frog is chased or caught by a grass snake.

Each female lays 1,000–2,000 eggs, deposited in a mass at the bottom of the water, the larger numbers being laid by the larger females. At first they are only about $\frac{1}{10}$ in. (2·5 mm.) in diameter, but the gelatinous covering absorbs so much water that they swell up to $\frac{1}{3}$ in. (8 mm.), become buoyant and float at the surface. Each of the jelly-spheres has a black centre, the egg proper, with a white spot on the lower side. In about four weeks the black centres develop into brown larvae, or tadpoles. Then, when the developing embryo has acquired an oval shape, it begins to spin on its axis, faster and faster, until it bursts from its jelly capsule. Having escaped, each larva clings to the remains of the jelly mass by a pair of suckers on the underside of the head. At this stage the larvae have no indication of limbs, and their head, body and tail, like those of a fish, merge into one. Even the gills are as yet undeveloped, though what may be termed the buds of them are seen on the bars separating the slits behind the head on each side. These buds soon expand into gill-plumes through which the blood circulates, taking up oxygen from the water that passes between them. There is no mouth at first, but this soon appears and horny plates on the jaws enable the tadpole to crop soft vegetable matter, upon which it subsists chiefly. Later on, the gill-plumes are hidden by a fold of skin growing over them.

Ultimately, the limbs appear. Though all four develop simultaneously, the hind pair are visible first, because the fore-limbs are at first hidden by the flap which grew over the gills. As the gill-plumes disappear lungs are developed and the animal changes from a fish-like water breather to an air breather, in preparation for a life on the land. By the time all the legs are well grown the form of the tadpole has changed to that of a frog, except for a long tail. This is later *absorbed*, not shed, and finally the hind end of the body is rounded off and there is nothing left to indicate that it once ended in a tail. The complete metamorphosis takes about ten weeks, and at the end of May or beginning of June the young frogs are ready to leave the water, although in mountain regions there may be some over-wintering of tadpoles, partly due to a late spawning and partly to cold weather inhibiting growth of the tadpoles.

When the change is complete the young frogs, their legs sufficiently firm to enable them to indulge in hopping exercises, still venture no farther than the very shallow water at the margin of the pond, where they can walk partially submerged, or at most take short excursions on land. Then comes a heavy rain and all have the impulse to leave the pond for the wet grass. The moment for their dispersal has arrived. At such times the ground may be littered with tiny frogs for 100 yd. (91 m.) or more, and it is fairly certain that it is this sudden appearance of many tiny frogs immediately after a heavy downpour that explains at least partially the stories of 'rains of frogs'.

At the time the young frogs disperse they measure $\frac{1}{2}$ in. (13 mm.) or so, and by the time they go into hibernation they are still less than 1 in. (25 mm.) in length. By the following autumn they will have doubled this length. They become sexually mature at three years; and the greatest age recorded, in captivity, is twelve years. Maximum lengths attained, in the south of England, are 3 in. (76 mm.) for the male and nearly 3½ in. (89 mm.) for the female. In the north of Scotland nearly 3¼ in. (82 mm.) was recorded for a male and just over 3¾ in. (95 mm.) for a female.

The common frog eats insects, slugs and worms. It has numerous enemies including herons, hawks, crows, ducks, gulls and owls, as well as otters, hedgehogs, stoats, weasels, badgers and rats. Its chief enemy is, of course, the grass snake.

Hibernation begins, according to the weather, in October or November. On mild days, frogs may leave their winter quarters for a brief spell, if they have hibernated on land. Many spend the winter in the mud at the bottom of a pond, but others find shelter in holes in banks, in crevices between rocks, and in other such places.

The common frog is widely distributed all over Britain; in Ireland, where it was introduced early in the eighteenth century, it occurs now in certain parts only. It is absent also from some Scottish islands. Although still called 'common' this frog is becoming rare. It has been much collected for dissection in biology classes and this, with the general lowering of the water table throughout the country, the filling in of ponds and, almost certainly, the use of insecticides, has caused a marked reduction in the numbers of frogs.

The Edible Frog

Family RANIDAE   *Rana esculenta*

The common frog is the only species native to Britain, but two Continental species, the edible frog and the marsh frog, have been introduced into the eastern and southern counties of England. The first to be introduced was the edible frog. In the early part of the nineteenth century, about 1,500 specimens from France and Belgium were turned loose in the Fens. They are, however, no longer plentiful there, though they occur locally in various parts of Norfolk. These, and a colony introduced into Cambridgeshire at about the same time, appear to have largely died out by 1914, and fresh importations from the Continent were liberated in recent years in Hampshire, Surrey, Oxfordshire and Bedfordshire.

The edible frog reaches a rather larger size than the common frog. It is usually without the dark patch extending from the eye to the shoulder, and the markings of the body, especially the bright yellow and black marblings of the hinder parts, are darker and bolder. Fully grown examples measure 2½–4 in. (64–101 mm.), head and body; the females are larger than the males. The head is more slender than in the common frog, and the brown ear drum is two-thirds of the diameter of the eye. The teeth on the palate form two oblique lines, and there is a pair of glandular folds behind the eye. The ground colour of the upper parts ranges from dull brown through olive to bright green, with dark brown or blackish spots on the back and larger patches of similar tint on the limbs. There is usually a line, running down the middle of the back from the muzzle to the hinder extremity, light green, yellow or golden in colour. The back of the thigh is always marbled with black and green or yellow. Though the thigh of the common frog is barred or blotched, it never bears these additional spots. The colour varies widely, according to the environment. Where exposed to sunlight, the frog is usually green, but on emergence from hibernation tends to be brown. Also, it is generally much brighter where vegetation is light than in dark swamps with sombre vegetation.

It is the habit of the edible frog to bask in the sun at the surface of the water and wait until its food comes within range of its extensible tongue. It does not study its prey like the common frog but jumps at it, even leaping up to grab butterflies. Its food consists of insects, especially beetles caught on land, and worms which are crammed into the mouth with the hands. It does not leave the water much, and quickly takes to it again when disturbed. In every way it gives the impression of being more active and agile than the common frog, although it is much more timid.

The most distinctive feature is restricted to the male. At the hinder angle of the mouth, just below the ear, are external vocal sacs which

16

can be distended with air to the size of large peas, giving the frog a quaint appearance. These act as resonators as well as increasing the volume of air across the vocal chords. The males continue to 'sing' after the breeding season is past, particularly on warm moonlight nights when a chorus of several hundreds may be heard for over a mile. Variations in the volume of sound depend on the amount of air in the vocal sacs.

Adults hibernate in mud at the bottom of ponds or at the water's edge, the young ones hibernating on land, under stones or logs, or in crevices. Emergence from hibernation is in April, but mating does not take place until May or June.

The life-history of the edible frog, from the egg to the loss of the tadpole's tail, follows much the same course as that of the common frog. There are, however, the following differences. The eggs are smaller but not more numerous. There have been records of one female producing from 5,000–10,000 eggs, but in this country the number has been found not to exceed 2,000. They are laid over a period of twenty-four to forty-eight hours, in groups of about 250, not in large masses. The tadpoles leave the eggs after seven to ten days, but do not metamorphose until three to four months later. Fully grown tadpoles are noticeably large, about $2\frac{1}{2}$ in. (63 mm.) long, of which more than $1\frac{1}{2}$ in. (38 mm.) is tail. The young frogs do not wander like those of the common frog, but remain in the vicinity of their birthplace. They become sexually mature at two years of age, but reach full size in their fourth and fifth years.

This is the frog of which the hind-legs are cooked and eaten. The edible frog is found all over Europe and in northern Asia.

The Marsh Frog

Family RANIDAE   *Rana ridibunda*

The marsh frog was first introduced in 1935, when twelve specimens were liberated in a pond on the edge of the Romney Marshes. Since then, they have penetrated the dykes and canals of the marshes, including the Walland Marsh. It is not certain whether the marsh frog and the edible frog are distinct species or races of a single species, so we can be more concerned with the differences between them, for in appearance they have much in common.

The marsh frog is as variable in colour as the edible frog. It is usually light or dark brown to greyish-brown above, suffused with green, either over the front half of the body or the whole of it. There are irregular black spots over the back and the limbs are barred with dark green, the hinder parts of the thighs being marbled with dark green and white or yellow. The lower parts are white, sometimes speckled with black or dark green. The tympanum is light brown, the iris gold with black markings, and, as a rule, there is no stripe in the mid-line of the back. When not exposed to the sun the body tends to be brown or even black. It differs from the edible frog in its longer snout and longer hind-limbs and there are more numerous warts on the skin. The marsh frogs do not assemble in large numbers to breed and there is a marked antagonism between breeding males. One having taken up position to call will chase away others that come too near.

The food consists largely of aquatic insects and freshwater shrimps, even small fish being taken occasionally. The frog has the habit of basking in the sun or of climbing on to the bank for the same purpose, whence it leaps into the water if approached.

Its natural range is Europe and western Asia.

The Common Toad

Family BUFONIDAE   *Bufo bufo*

Most people have little difficulty in distinguishing between a frog and a toad, but the two are sufficiently alike to cause momentary confusion. A toad has a flatter back than the common frog, and the hind-legs are not so long in proportion to the body, only slightly exceeding the length of head and body. It appears more solidly built, with its broader head, shorter limbs and a heavier, more grovelling movement than the vaulting frog. Instead of the moist and shining, bright-coloured coat of the frog, it has a dry, dull, pimply skin so strongly resembling the earth that it is frequently overlooked as a lifeless clod. This resemblance to its usual background is increased by the toad's habit of squatting motionless for hours. A toad is too heavy to leap, but progresses by very short jumps on all four feet, which give the impression of being accomplished only by a great effort. When in search of food it walks, yet it swims well and climbs surprisingly well even if its actions appear laborious. Toads have been known to have habitual resting places in old birds' nests a foot or two from the ground in a privet hedge. They often climb banks or the faces of quarries on migration.

The colour varies a good deal according to the nature of the soil upon which the toad is living. It is usually some tint of brown or grey, but the brown may be almost red in a sand-pit, and in other places it may be a rich or a dirty brown. In other situations it may be light grey, perhaps with an orange tinge, or it may be a sooty hue which is almost black. As a toad is active mainly in the evening and at night, any of these tints serve to render it inconspicuous in a dim light. The eyes are golden or a coppery-red in colour. The underside is whitish, the white being qualified always with an admixture of yellow, brown or red, sometimes spotted with black.

The skin is covered with wrinkles and irregular and conspicuous warts, the largest of which bear central spines or tubercles. The pattern produced helps materially the resemblance to a clod of earth. The largest gland, the parotid, is seen as an elongated, porous swelling behind the eye. When a toad is gripped, or even in circumstances of alarm without being touched, its glands secrete an acrid and offensive fluid obnoxious to those animals not normally predators. Experience teaches such enemies to leave toads alone. A dog merely touching a toad with its lips salivates copiously and this may be followed by nausea and vomiting.

The skin is shed several times in the course of the summer. Distending the body and humping the back causes it to split along a thin line known as the raphe, extending down the middle of the back. The skin slides off each side and that of the head is scraped off with the fore-limbs and crammed into the mouth and swallowed. The rest peels, assisted by contortions of the body and limbs, and is also swallowed. The new skin is moist but soon dries.

In head and body length the males measure about $2\frac{1}{2}$ in. (63 mm.), and females 1 in. (25 mm.) longer. Occasionally much larger examples

are met and these monsters are almost certainly females. The female
is without voice, and the male has no vocal sacs, internal or external, so
that the best he can accomplish is a subdued croak, used when mating.
In calling to the female before pairing, he uses a short staccato croak
which has been described as being like the distant bark of a small dog.
He also uses a short chirping note to threaten another male.

The male develops special grasping pads at pairing time on the palm
and three inner fingers.

After the breeding season toads wander away from the water, and
distribute themselves over field, hedgerow, wood and garden, wherever
there is an abundance of insect life, for the quantity of food each toad
consumes is enormous. Beetles and ants form a large part of the diet,
to which may be added a variety of other insects and their larvae, woodlice, earthworms and snails. In fact, a toad will eat almost anything of
appropriate size that comes its way, but its prey must be moving. A
well-known habit is for a toad to station itself outside a beehive at dusk,
snapping up the late homing bees one by one. It has been seen doing
the same at a wild bees' nest, and there has been one record of a nest
of wasps being treated in the same way. It would seem that this method
of feeding is very much an individual matter, for under test toads were
seen to take bees until stung in the mouth, after which they tended to
avoid them. Small snails are swallowed whole, large ones are broken in
the mouth first. Other additions to the diet are young newts, frogs and
toads, young slow-worms and snakes, the hands often being used to
cram such prey into the mouth. The size of a toad's appetite may be
gauged from one whose stomach contained 363 ants. Small prey, such
as insects, are caught with the tongue, in the manner described for a
frog (see p. 202).

A toad spends the hotter part of the day concealed under the lower
foliage of plants, under a stone or log, in a crevice or cavity in the ground
scooped out with the hind-legs. As many nocturnal insects seek similar
situations in the day-time, it has no difficulty in finding food even during
the day. A shower of rain, which brings snails and worms out, will also
tempt a toad into the open.

The toad has the homing faculty well developed, home being the
hollow it has scooped out, the crevice in rock or the cavity under a root

or a stone it has chosen. In the evening it sets out foraging, and may travel long distances, but before morning it is back snugly in its form, occupying the same place day after day for many months. Toads tested for their homing abilities have found their way back, even though it may take several days to do so.

A similar sense of locality is shown in the choice of ponds for breeding. The migration to the breeding ponds is more spectacular than that of frogs. In early spring scores of toads may be seen converging upon a particular pond, perhaps passing other pieces of water, that would seem quite suitable for their purpose. It is highly likely that in such cases the toads are making their way back to the identical pond in which they developed in the way that migrant birds will find their way back to build their nests in the copse or hedgerow where they were hatched. There is similar evidence to that given for frogs (q.v.) that toads must use celestial navigation in both migrations and homing.

Migrations may begin in early February, but the main stream is not seen until the second half of March or the beginning of April, in southern England, and up to the end of April in Scotland. The first signs are, all too often, the crushed corpses on the roads. These roads would appear to have been driven, very often, across traditional migration routes, judging by the large numbers of toads killed in some localities. The migration may occupy a week or more, or may be concluded within the space of twenty-four hours. Such is the determination to travel in a direct line that obstacles, such as stone walls built across the line of advance, will be laboriously climbed. During the migrations, also, no notice is taken of predators, or of human beings, and a heavy mortality often results. Some unpaired males may arrive first at the pond, but many reach the water already on the backs of the females.

Toads spawn in deeper water than frogs, and their eggs are often difficult to locate. The small, black eggs of the toad differ from those of the frog in being laid in series of three, or sometimes four, but within a few days they become arranged in a double row embedded in a gelatinous string 7–10 ft. (2–3 m.) in length, which may be stretched up to 15 ft. (4·5 m.). As in frogs, the gelatinous coating around the eggs absorbs water to swell to three times its original size. The number of eggs in a string may be 3,000–4,000, or more. The strings are wound about the stems of water-weeds by the haphazard movements of the parents. The small black larvae hatch out in ten to twelve days, the jelly of the string having by then been partially dissolved. For the first few days the tadpoles cling to the egg-strings, then hang tail downwards from the undersides of leaves. They go through similar stages to those of the frog tadpole, and become small tail-less toads, a little more than ½ in. (12·5 mm.) long in eleven to twelve weeks. Although sexual maturity is attained after the fourth year it is five or six years before the males reach full size and possibly longer for the females. Toads are long-lived. In captivity, an age of forty years or more has been attained. In old age they frequently succumb to the attacks of flesh-eating flies (*Lucilia bufonivora*), the eggs of which are deposited on the backs of the toads, the small maggots hatching from them entering the toads' nostrils. Natural enemies include hedgehogs, stoats and weasels, as well as rats. Crows and magpies attack them on their breeding migrations. Some grass snakes will eat toads; others will not. Sometimes a toad's reaction to the approach of a grass snake is to blow itself up, when its inflated body is seen supported upon four stiff legs. In this attitude the snake is unable to obtain a grip on the toad to swallow it.

Some years ago naturalists were puzzled by finding masses of whitish jelly on the ground. This was popularly known as star-slime, or 'rot of the stars', supposedly having something to do with shooting stars. The masses proved to be the linings of oviducts of toads and frogs, presumably devoured by birds and regurgitated; these swelled up when exposed to moisture.

The common toad is found all over England, Wales and Scotland, but Ireland appears never to have had it, in spite of the legend that St Patrick banished it with the snakes. Outside Britain it ranges over Europe and much of Asia.

The Natterjack Toad

Family BUFONIDAE   *Bufo calamita*

Although in general appearance the natterjack toad resembles the common toad, it shows many small differences over and above its smaller size and its colouration. Its legs are actually shorter and also proportionately so. Its most distinctive feature is, however, the narrow yellow line that runs along the centre of the head and back, which has given rise to one of its local names, 'golden-back'. 'Running toad' is the good descriptive name by which it is known in the Fens, for owing to the shortness of the hind-limbs the natterjack does not hop. It runs for a short distance, then stops for a while, and runs on again. Moreover, it is a poor swimmer.

The maximum length of head and body is under 3 in. (76 mm.), and there is no marked difference in size between the sexes. In the breeding season the male develops nuptial pads on his first three fingers, and he has a large internal vocal sac, which, when in use, causes a great bulging of his bluish or violet throat. The ground colour of the body is greyish or pale yellowish-brown tending to olive, with clouding and distinct spots of a darker brown, reddish, yellowish or greenish hue. The underside is yellowish-white with black or very dark green spots, and the legs are barred with black. The prominent eyes are greenish-yellow, veined with black, and the long parotid gland behind the eye is smaller than in the common toad. The male, in the breeding season, can also be distinguished by its much stronger fore-limbs.

The natterjack breeds later than the common species, the pairing not beginning until the end of March or April and being spread over May and June. In cold summers it may extend to the beginning of August. The natterjack appears to be less tied to a spawning ground than the common toad, and the locality, when chosen, is advertised by the rattling noise of the males, a loud trilling croak continued for a few seconds at a time, and of sufficient power to be heard a long distance away. Mating, which is carried out in shallow water, appears to take place at night, after which the natterjack leaves the water to hide in reeds or under stones. This seems to vary somewhat, though, for in the north of England it goes on night and day. The egg-strings are short as compared with those of the common toad, being only 5–6 ft. (1·3–1·5 m.) in length. The blackish tadpoles are 1 in. (25 mm.) long when fully grown, but the young toads are less than ½ in. (12·5 mm.) long when they leave the water, and under 1 in. (25 mm.) long by the autumn. The development into tail-less toads takes six to eight weeks. In their second year they measure 1–1½ in. (25–40 mm.), and when they become mature, between the fourth and fifth years, they are only $1\frac{1}{2}$–2 in. (35–50 mm.) long.

The natterjack feeds on insects and worms, spiders, woodlice, slugs and snails, and though its activities are mainly nocturnal, it may be seen running about in full sunshine. When alarmed it plays the same trick as the common toad, blowing itself up, but because of its short legs the posture is not the same. It is also said to spread itself out flat on the ground, 'feigning dead'. The secretion from its glands when it is annoyed is said to smell 'of gun-powder or india-rubber'.

In hibernation the natterjack buries itself one to two feet in the ground. It has also been known to climb up steep walls of sand to winter in the nesting burrows of sand martins.

The natterjack is plentiful in some English localities and is often found in sandy coastal areas, but it appears to be somewhat migratory. Many places where it has been recorded one year fail to yield even a solitary specimen the next year. It is also found in south-west Ireland, and ranges over much of western and central Europe.

Sir Joseph Banks first called attention to it as a British species in the account published in Pennant's *British Zoology* (1776). Part of this note is worth quoting: 'This species frequents dry and sandy places and in Lincolnshire is known as the Natter Jack. It never leaps, neither does it crawl, but its motion is liker to running. Several are found commonly together, and like others of the genus, they appear in the evenings.'

BOOKS FOR FURTHER READING

Crowcroft, P. *The Life of the Shrew* (Reinhardt)

Godfrey, G. and Crowcroft, P. *The Life of the Mole* (Museum Press)

Lawrence, M. J. and Brown, R. W. *Mammals of Britain: Their Tracks, Trails and Signs* (Blandford)

Matthews, L. Harrison. *British Mammals* (Collins' New Naturalist Series)

Neal, E. *The Badger* (Collins' New Naturalist Monograph)

Shorten, M. *Squirrels* (Collins' New Naturalist Monograph)

Smith, Malcolm A. *The British Amphibians and Reptiles* (Collins' New Naturalist Series)

Southern, H. N. *Handbook of British Mammals* (Blackwell)

Taylor Page, F. J. *Field Guide to British Deer* (Mammal Society of British Isles)

Thompson, H. V. and Worden, A. N. *The Rabbit* (Collins' New Naturalist Monograph)

INDEX

Adder, xv, 190, *Pls.* 33, 35
Albino Mice, 88
American Mink, 147, *Pl.* 30
Amphibians, xv, xxiii, 194–215
ANGUIDAE, 179
Anguis fragilis, 179
Apodemus flavicollis, 94
— *sylvaticus*, 89
— — *butei*, 94
— — *fridariensis*, 94
— — *hebridensis*, 94
— — *hirtensis*, 94
Armies of Rats, 96, 102
Artiodactyla, xxi
Arvicola terrestris, 81
— — *brigantium*, 81
— — *reta*, 82
Atlantic Seal, xvi, 157, *Pl.* 39
Aurochs, xvii

Back-boned Animals—*see* Vertebrates
Badger, 126, *Pls.* 20, 23
Bank Vole, 75, *Pl.* 12
Banks, Sir Joseph, 215
Barbastelle, 54, *Pl.* 11
Barbastella barbastellus, 54
Barn Rat, 100
Barry, the Rev. George, 80
Bat, Bechstein's, 43, *Pl.* 8
—, Common, 52, *Pl.* 4
—, Daubenton's, 45, *Pl.* 10
—, Great, 50, *Pl.* 14
—, Grey long-eared, 57
—, Hairy-winged, 49
—, Horseshoe, Greater, 34, 39, *Pl.* 5
—, —, Lesser, 34, 39, *Pl.* 5
—, Leisler's, 49, *Pl.* 11
—, Long-eared, 55, *Pl.* 14
—, Mouse-eared, 44, *Pl.* 6

Bat—(*contd.*)
—, Natterer's, 42, *Pl.* 7
—, Noctule, 34, 47, 50, *Pl.* 14
—, Red-grey, 42
—, Serotine, 34, 47, *Pl.* 10
—, Water, 45
—, Whiskered, 40, *Pl.* 6
—, White's, 50
Bat's Echo-location, 29
Bats, xvii, xxv, 28–57
Beale—*see* Weasel, 144
Bear, Brown, xvii
Beasts of Prey, xvii
Beaver, xvii
Bechstein's Bat, 43, *Pl.* 8
Birds, xi, xv
Blind-worm, 179
Black Rat, 97, *Pl.* 16
Blackmore, Michael, 44
Blue Hare, 69
Blue-spotted Slow-worm, 182
Boar, Wild, xvii
BOVIDAE, 172
Brown Bear, xvii
— Hare, 64, *Pl.* 15
— Rat, 97, 99, *Pl.* 16
Bufo bufo, 209
— *calamita*, 213
BUFONIDAE, 209, 213
Bute Field Mouse, 94

Cairn-building of Field Mouse, 93
Cane Weasels, 144
CANIDAE, 118, 124
Canis familiaris, 124
Capra hircus, 172
Capreolus capreolus, 165
CAPROMYIDAE, 116
Carnivora, xvii

Cat, Feral, xvi, 148
—, Marten, 135
—, Wild, 151, *Pl.* 25
CERVIDAE, 161, 165, 168, 170, 171, 172
Cervus elaphus, 161
— *nippon*, 170
Cetacea, xvi, xvii
Charming, 123, 139, 143
Chinese Muntjac, xvi, 171
— Water Deer, 172, *Pl.* 34
Chiroptera, xvii, 28–57
Clethrionomys glareolus, 75
Cloven-hoofed Animals, xxi, 161–173
Colour change, xxiii, 3, 138, 141, 144
COLUBRIDAE, 183, 188
Common Bat, 52, *Pl.* 4
— Frog, xv, 200, 206, *Pl.* 37
— Lizard, xv, 174
— Newt, xv, 198
— Rat, xv, 97, 99, *Pl.* 16
— Seal, 155, *Pl.* 39
— Shrew, 11, 17, *Pl.* 1
— Toad, xv, 209, *Pl.* 40
Conies, 58
Continental Water Vole, 82
Coronella austriaca, 188
Coypu, xvi, 116, *Pl.* 30
Crested Newt, xv, 195, *Pl.* 36
Crowcroft, Peter, xii

Dama dama, 168
Daubenton's Bat, 45, *Pl.* 10
Dead-adder, 179
Deer, Fallow, xvi, 168, *Pl.* 29
—, Red, 161, *Pls.* 28, 31
—, Roe, 165, *Pls.* 29, 31
Delayed Fertilisation, 37, 53
Delayed Implantation, 129, 134, 136, 140, 166
Derrymouse, 73
'Devil's Hoofmarks', 92
Dog, Feral, xvi, 124
Dolphins, xvi, xvii
Dormouse, 71, *Pls.* 9, 21
—, Edible, 74, *Pl.* 9
—, Fat, 74, *Pl.* 9
—, Squirrel-tailed, 74

Dorymouse, 73
Dozing-mouse, 73
Drey, Squirrel's, 107
Duplicidentata, xx, 64

Echo-location of the Bat, 29
Edible Dormouse, 74, *Pl.* 9
— Frog, xv, 206, *Pl.* 37
Eft, 198
Eptesicus serotinus, 47
ERINACEIDAE, 20
Erinaceus europaeus, 20
European Mink, 148
Evat, 198

Fair Isle Field Mouse, 94
Fallow Deer, xvi, 168, *Pl.* 29
'Fancy' Mice, 88
Fat Dormouse, 74, *Pl.* 9
FELIDAE, 148, 151
Felis catus, 148
— *silvestris grampia*, 151
Feral Cat, xvi, 148
—, Dog, xvi, 124
Field Mouse, Bute, 94
— —, Fair Isle, 94
— —, Hebridean, 94
— —, Long-tailed, 89, *Pls.* 13, 21
— —, St Kilda, 94
— —, Shetland, 94
Field Vole, 77, *Pl.* 12
Fishes, xv
Fitchew, 145
Flesh Eaters, 118–160
Flying Mammals, 28–57
'Fortress', Mole's, 5
Fossils, xix
Foul-marten, 145
Foumart, 145
Fox, 118, *Pls.* 20, 23, 24
Frog, Common, xv, 200, 206, *Pl.* 37
—, Edible, xv, 206, *Pl.* 37
—, Marsh, xv, 206, 208, *Pl.* 38
Furze evvet, 177

Galambos, R., 30
GLIRIDAE, 71, 74

Glis glis, 74
Gnawing Animals, xvii, 58–117
Goat, 172
Godfrey, Gillian, xii
Grass Snake, xv, 183, *Pl.* 35
Great Bat, 50, *Pl.* 14
— Newt, 196
Greater Horseshoe Bat, 34, 39, *Pl.* 5
Green Lizard, 179
Grey Long-eared Bat, 57
— Seal, xvi, 157, *Pl.* 39
— Squirrel, xv, 109, Pl. 22
Griffin, D. R., 30

Hairy-winged Bat, 49
Hanoverian Rat, 100
Harbor Seal, 155
Hare, Blue, 69
—, Brown, 64, *Pl.* 15
—, Irish, 70
—, Scottish, 69
—, Variable, 69
—, White, 69
Hares, xvii, xx, xxi
Harriman, 177
Hartridge, Professor, 29
Harvest Mouse, 83, *Pls.* 13, 19
Hebridean Field Mouse, 94
Hedgehog, xvii, xxii, 20, *Pls.* 1, 2, 3
Hedgepig, 20
Hibernating Gland, 22
Horseshoe Bat, 28, 42
— —, Greater, 34, 39, *Pl.* 5
— —, Lesser, 34, 39, *Pl.* 5
House Mouse, xv, 86, *Pl.* 22
Hummels, 162
Hydropotes inermis, 172

Insect Eaters, 1–27
Insectivora, xvii
Irish Hare, 70
Islay Stoat, 140

Jug, Squaggy—*see* Drey, Squirrel's
Jurassic Period, xi

Kine, 144
Kolb, A., 45

Lacerta agilis, 178
— *muralis*, 179
— *viridis*, 179
— *vivipara*, 174
LACERTIDAE, 174, 178
Lagomorpha, xvii, xxi, 64
Lake Monsters, 134
Le Court, 5
Least Weasel, 143
Leisler's Bat, 49, *Pl.* 11
LEPORIDAE, 58, 64, 69, 70
Lepus europaeus, 64
— *timidus*, 69
— — *hibernicus*, 70
— — *scoticus*, 69
Lesser Horseshoe Bat, 34, 39, *Pl.* 5
— Shrew, 16, 83
Leutscher, Alfred, 92
Lizard, Common, xv, 174
—, Green, 179
—, Sand, xv, 178, *Pls.* 32, 35
—, Viviparous, xv
—, Wall, 179
Lizards, xv, 174–182
Long-eared Bat, 55, *Pl.* 14
— —, Grey, 57
Long-tailed Field Mouse, 89, *Pls.* 13, 21
Lutra lutra, 131

Mammals, xi, xv
—, Flying, 28–57
—, Marine, xvi
Marsh Frog, xv, 206, 208, *Pl.* 38
Marten Cat, 135
Martes martes, 135
Matheson, Colin, 68
Meles meles, 126
Mice, Albino, 88
—, 'Fancy', 88
Micromys minutus soricinus, 83
Microtus agrestis, 77
— *arvalis orcadensis*, 80
Millais, J. G., 80
Mink, American, 147, *Pl.* 30
—, European, 148
Moffat, C. B., 51

Mole, xii, xvii, 1, *Pl.* 2
Mole-heaves, 5
Mole's 'Fortress', 5
Montagu, George, 86
Mouse, Field, Bute, 94
— —, Fair Isle, 94
— —, Hebridean, 94
— —, Long-tailed, 89, *Pls.* 13, 21
— —, St Kilda, 94
— —, Shetland, 94
—, Harvest, 83, *Pls.* 13, 19
—, House, xv, 86, *Pl.* 22
—, St Kilda, xv, 89
—, Wood, 89, *Pls.* 13, 21
—, Yellow-necked, 94
Mouse-eared Bat, 44, *Pl.* 6
Mouse-killer, 144
Muntiacus reevesi, 171
MURIDAE, 75, 77, 80, 81, 83, 86, 89,
  94, 97, 99
Mus muralis, 89
— *musculus*, 86
Muscardinus avellanarius, 71
Mustela erminea, 138
— — *hibernica*, 140
— — *ricinae*, 140
— *lutreola*, 148
— *nivalis*, 140, 142
— *rixosa*, 143
— *vison*, 147
MUSTELIDAE, 126, 131, 135, 138, 142,
  145
Myocastor coypus, 116
Myotis bechsteini, 43
— *daubentoni*, 45
— *myotis*, 44
— *mystacinus*, 40
— *nattereri*, 42

Natrix natrix, 183
Natterer's Bat, 42, *Pl.* 7
Natterjack Toad, xv, 213, *Pl.* 40
Neomys fodiens, 17
Newt, Common, xv, 198
—, Crested, xv, 194, *Pl.* 36
—, Great, 196
—, Palmate, xv, 199, *Pl.* 38

Newt—(*contd.*)
—, Smooth, 198, *Pls.* 36, 38
—, Spotted, 198
—, Warty, 196
Noctule, Bat, 34, 47, 50, *Pl.* 14
Norway Rat, 97, 100
Nose-leaves, 29
Nutria, xvi, 116
Nyctalus altivolans, 50
— *leisleri*, 49
— *noctula*, 50

Oldham, Charles, 35
Orkney Vole, 80, *Pl.* 18
Oryctolagus cuniculus, 58
Otter, 131, *Pl.* 26
Ox, Wild, Short-horned, xvii

Pallas, Peter, 86
Palmate Newt, xv, 199, *Pl.* 38
Pennant, 83, 86, 152
Permian Period, xi
Phoca vitulina, 155
PHOCIDAE, 155, 157
Pine Marten, 135, 145, *Pl.* 30
PINNIPEDIA, 153
Pipistrelle, 40, 52, *Pl.* 4
Pipistrellus pipistrellus, 52
Plecotus auritus, 55
— *austriacus*, 57
Pleistocene remains, xvi
Polecat, 145, *Pl.* 25
Porpoises, xvi
Pricket, 169
Proboscidea, xxi
Putorius putorius, 145
Pygmy Shrew, 16, 83

Rabbit, xv, xvii, 58, *Pl.* 4
Rana esculenta, 206
— *ridibunda*, 208
— *temporaria*, 200
RANIDAE, 200, 206, 208
Ransome, H. D., 36
Rat, Barn, 100
—, Black, 97, *Pl.* 16
—, Brown, 97, 99, *Pl.* 16

Rat—(*contd.*)
—, Common, xv, 97, 99, *Pl.* 16
—, Hanoverian, 100
—, House, 97
—, Norway, 97, 100
—, Roof, 97
—, Sewer, 97
—, Ship, xv, 97, *Pl.* 16
—, Water, 81
Rats Carrying Eggs, 103
Rattle-mouse, 49
Rattus rattus, 97
— — *alexandrinus*, 99
— — *frugivorus*, 99
— — *rattus*, 99
— *norvegicus*, 99
Red Deer, 161, *Pls.* 28, 31
— Squirrel, 105, 109, *Pl.* 17
Red-grey Bat, 42
Refection, 63, 67
Reid, Capt. Mayne, 149
Reptiles, xi, xv
Resorption of Embryos, 7, 13, 59, 68
Ressel, 144
RHINOLOPHIDAE, 34, 39
Rhinolophus ferrum-equinum, 34
— *hipposideros*, 39
Rodentia, xvii, xxi
Roe Deer, 165, *Pls.* 29, 31
— Rings, 166
Rothschild, Walter, 74

St John, C., 152
St Kilda Field Mouse, 94
St Kilda Mouse, xv, 89
SALAMANDRIDAE, 194, 198, 199
Sand Lizard, xv, 178, *Pls.* 32, 35
SCIURIDAE, 105, 109
Sciurus carolinensis, 109
— *vulgaris leucourus*, 105
Scottish Hare, 69
Seal, Atlantic, xvi, 157, *Pl.* 39
—, Common, xvi, 155, *Pl.* 39
—, Grey, xvi, 157, *Pl.* 39
—, Harbor, 155
Seals, xvi, 153–160
Self-anointing in Hedgehogs, 21

Serotine Bat, 34, 47, *Pl.* 10
Sewer Rat, 97
Shetland Field Mouse, 94
Ship Rat, xv, 97, *Pl.* 16
Short-horned Wild Ox, xvii
Short-tailed Vole, 77, *Pl.* 12
Shrew, Common, 11, 17, *Pl.* 1
—, Lesser, 16, 83
—, Pygmy, 16, 83
—, Water, 17, *Pls.* 2, 3
Shrews, xvii
Sika, xvi, 170, *Pls.* 28, 34
Simplicidentata, xx, 64
Simpson, Dr G. G., xix
Skeleton, xvii
Sleepmouse, 73
Slow-worm, xv, 179, *Pl.* 32
—, Blue-spotted, 182
Smooth Newt, 198, *Pls.* 36, 38
— Snake, xv, 188, *Pl.* 33
Snake, Grass, xv, 183, *Pl.* 35
—, Smooth, xv, 188, *Pl.* 33
Snakes, xv, 183–193
Sorex araneus, 11
— — *granti*, 16
— *granti*, 16
— *minutus*, 16
SORICIDAE, 11, 16, 17
Southern, H. N., 60, 61, 62
Spotted Newt, 198
Squaggy Jug, *see* Drey, Squirrel's
Squirrel, Drey of, 107
—, Grey, xv, 109, *Pl.* 22
—, Red, 105, 109, *Pl.* 17
Squirrel-tailed Dormouse, 74
'Stab' of a Rabbit, 62
Stoat, 138, *Pl.* 27
Sweet-mart, 145
Swift, 177

Talpa europaea, 1
TALPIDAE, 1
Teeth, xviii
Theromorpha, xi
Thompson, Harry, 68
Toad, Common, xv, 209, *Pl.* 40
—, Natterjack, xv, 213, *Pl.* 40

Tod, 118
Tragus, 29
Triturus cristatus, 194
— *helveticus*, 199
— *vulgaris*, 198
Tylopoda, xxi

Ungulata, xxi
Urchin, 20

Variable Hare, 69
Vertebrates, xv
VESPERTILIONIDAE, 30, 31, 33, 40, 42, 43, 44, 45, 47, 49, 50, 52, 54, 55, 57
Viper, xv, 190, *Pls.* 33, 35
Vipera berus, 190
VIPERIDAE, 190
Viviparous Lizard, xv
Vole, Bank, 75, *Pl.* 12
—, Field, 77, *Pl.* 12
—, Orkney, 80, *Pl.* 18
—, Short-tailed, 77, *Pl.* 12
—, Water, 81, *Pl.* 18
—, —, Continental, 82
Voles, xv
Vulpes vulpes, 118

Wall Lizard, 179
Warty Newt, 196
Water Bat, 45
'Water Rat', 81
Water Shrew, 17, *Pl.* 3
— Vole, 81, *Pl.* 18
Weasel, 140, 142, *Pls.* 17, 27
—, Cane, 144
—, Least, 143
Whales, xvi, xvii
Whiskered Bat, 40, *Pl.* 6
White, Gilbert, 46, 50, 83, 86, 144
White Hare, 69
White's Bat, 50
Whitethroat, 144
Whittret, 144
Wild Boar, xvii
— Cat, 151, *Pl.* 25
Wolf, xvii
Wood Mouse, 89, *Pls.* 13, 21
Woodruffe-Peacock, 68

Yellow-necked Mouse, 94
Yugoslavia, Armies of Mice, 95